Geometry and Physics

Jürgen Jost

Geometry and Physics

 Springer

Jürgen Jost

Max Planck Institute
for Mathematics in the Sciences
Inselstrasse 22
4103 Leipzig
Germany
jjost@mis.mpg.de

ISBN 978-3-642-00540-4   e-ISBN 978-3-642-00541-1
DOI 10.1007/978-3-642-00541-1
Springer Heidelberg Dordrecht London New York

Library of Congress Control Number: 2009934053

Mathematics Subject Classification (2000): 51P05, 53-02, 53Z05, 53C05, 53C21, 53C50, 53C80, 58C50, 49S05, 81T13, 81T30, 81T60, 70S05, 70S10, 70S15, 83C05

Cover design: WMX Design

Printed on acid-free paper

Springer is part of Springer Science+Business Media (www.springer.com)

Dedicated to Stephan Luckhaus,
with respect and gratitude for his critical mind

The aim of physics is to write down the Hamiltonian of the universe. The rest is mathematics.

Mathematics wants to discover and investigate universal structures. Which of them are realized in nature is left to physics.

Preface

Perhaps, this is a bad book. As a mathematician, you will not find a systematic theory with complete proofs, and, even worse, the standards of rigor established for mathematical writing will not always be maintained. As a physicist, you will not find coherent computational schemes for arriving at predictions.

Perhaps even worse, this book is seriously incomplete. Not only does it fall short of a coherent and complete theory of the physical forces, simply because such a theory does not yet exist, but it also leaves out many aspects of what is already known and established.

This book results from my fascination with the ideas of theoretical high energy physics that may offer us a glimpse at the ultimate layer of reality and with the mathematical concepts, in particular the geometric ones, underlying these ideas.

Mathematics has three main subfields: analysis, geometry and algebra. Analysis is about the continuum and limits, and in its modern form, it is concerned with quantitative estimates establishing the convergence of asymptotic expansions, infinite series, approximation schemes and, more abstractly, the existence of objects defined in infinite-dimensional spaces, by differential equations, variational principles, or other schemes. In fact, one of the fundamental differences between modern physics and mathematics is that physicists usually are satisfied with linearizations and formal expansions, whereas mathematicians should be concerned with the global, nonlinear aspects and prove the convergence of those asymptotic expansions. In this book, such analytical aspects are usually suppressed. Many results have been established through the dedicated effort of generations of mathematicians, in particular by those among them calling themselves mathematical physicists. A systematic presentation of those results would require a much longer book than the present one. Worse, in many cases, computations accepted in the physics literature remain at a formal level and have not yet been justified by such an analytical scheme. A particular issue is the relationship between Euclidean and Minkowski signatures. Clearly, relativity theory, and more generally, relativistic quantum field theory require us to work in Lorentzian spaces, that is, ones with an indefinite metric, and the corresponding partial differential equations are of hyperbolic type. The mathematical theory, however, is easier and much better established for Riemannian manifolds, that is, for spaces with positive definite metrics, and for elliptic partial differential equations.

In the physics literature, therefore, one often carries through the computations in the latter situation and appeals to a principle of analytic continuation, called Wick rotation, that formally extends the formulae to the Lorentzian case. The analytical justification of this principle is often doubtful, owing, for example, to the profound difference between nonlinear elliptic and hyperbolic partial differential equations. Again, this issue is not systematically addressed here.

Algebra is about the formalism of discrete objects satisfying certain axiomatic rules, and here there is much less conflict between mathematics and physics. In many instances, there is an alternative between an algebraic and a geometric approach. The present book is essentially about the latter, geometric, approach. Geometry is about qualitative, global structures, and it has been a remarkable trend in recent decades that some physicists, in particular those considering themselves as mathematical physicists (in contrast to the mathematicians using the same name who, as mentioned, are more concerned with the analytical aspects), have employed global geometric concepts with much success. At the same time, mathematicians working in geometry and algebra have realized that some of the physical concepts equip them with structures that are at the same time rich and tightly constrained and thereby afford powerful tools for probing old and new questions in global geometry.

The aim of the present book is to present some basic aspects of this powerful interplay between physics and geometry that should serve for a deeper understanding of either of them. We try to introduce the important concepts and ideas, but as mentioned, the present book neither is completely systematic nor analytically rigorous. In particular, we describe many mathematical concepts and structures, but for the proofs of the fundamental results, we usually refer to other sources. This keeps the book reasonably short and perhaps also aids its coherence. – For a much more systematic and comprehensive presentation of the fundamental theories of high-energy physics in mathematical terms, I wish to refer to the forthcoming 6-volume treatise [111] of my colleague Eberhard Zeidler.

As you will know, the fundamental problem of contemporary theoretical physics[1] is the unification of the physical forces in a single, encompassing, coherent "Theory of Everything". This focus on a single problem makes theoretical physics more coherent, and perhaps sometimes also more dynamic, than mathematics that traditionally is subdivided into many fields with their own themes and problems. In turn, however, mathematics seems to be more uniform in terms of methodological standards than physics, and so, among its practioners, there seems to be a greater sense of community and unity.

Returning to the physical forces, there are the electromagnetic, weak and strong interactions on one hand and gravity on the other. For the first three, quantum field theory and its extensions have developed a reasonably convincing, and also rather successful unified framework. The latter, gravity, however, more stubbornly resists such attempts at unification. Approaches to bridge this gap come from both sides. Superstring theory is the champion of the quantum camp, ever since the appearance

---

[1]More precisely, we are concerned here with *high-energy* theoretical physics. Other fields, like solid-state or statistical physics, have their own important problems.

of the monograph [50] of Green, Schwarz and Witten, but many people from the gravity camp seem unconvinced[2] and propose other schemes. Here, in particular Ashtekar's program should be mentioned (see e.g. [92]). The different approaches to quantum gravity are described and compared in [74]. A basic source of the difficulties that these two camps are having with each other is that quantum theory does not have an ontology, at least according to the majority view and in the hands of its practioners. It is solely concerned with systematic relations between observations, but not with any underlying reality, that is, with laws, but not with structures. General relativity, in contrast, is concerned with the structure of space–time. Its practioners often consider such ideas as extra dimensions, or worse, tunneling between parallel universes, that are readily proposed by string theorists, as too fanciful flights of the imagination, as some kind of condensed metaphysics, rather than as honest, experimentally verifiable physics. Mathematicians seem to have fewer difficulties with this, as they are concerned with structures that are typically believed to constitute some higher form of 'Platonic' reality than our everyday experience. In the present book, I approach things from the quantum rather than from the relativity side, not because of any commitment at a philosophical level, but rather because this at present offers the more exciting mathematical perspectives. However, this is not meant to deny that general relativity and its modern extensions also lead to deep mathematical structures and challenging mathematical problems.

While I have been trained as a mathematician and therefore naturally view things from a structural, mathematical rather than from a computational, physical perspective, nevertheless I often find the physicists' approach more insightful and more to the point than the mathematicians' one. Therefore, in this book, the two perspectives are relatively freely mixed, even though the mathematical one remains the dominant one. Hopefully, this will also serve to make the book accessible to people with either background. In particular, also the two topics, geometry and physics, are interwoven rather than separated. For instance, as a consequence, general relativity is discussed within the geometry part rather than the physics one, because within the structure of this book, it fits into the geometry chapter more naturally.

In any case, in mathematics, there is more of a tradition of explaining theoretical concepts, and good examples of mathematical exposition can provide the reader with conceptual insights instead of just a heap of formulae. Physicists seem to make fewer attempts in this direction. I have tried to follow the mathematical style in this regard.

I have assembled a representative (but perhaps personally biased) bibliography, but I have made no attempt at a systematic and comprehensive one. In the age of the Arxiv and googlescholar, such a scholarly enterprise seems to have lost its usefulness. In any case, I am more interested in the formal structure of the theory than in its historical development. Therefore, the (rather few) historical claims in this book should be taken with caution, as I have not checked the history systematically or carefully.

---

[2]For an eloquent criticism, see for example Penrose [85].

Acknowledgements

This book is based on various series of lectures that I have given in Leipzig over the years, and I am grateful for many people in the audiences for their questions, critical comments, and corrections. Many of these lectures took place within the framework of the International Max Planck Research School "Mathematics in the Sciences", and I wish to express my particular gratitude to its director, Stephan Luckhaus, for building up this wonderful opportunity to work with a group of talented and enthusiastic graduate students. The (almost) final assembly of the material was performed while I enjoyed the hospitality of the IHES in Bures-sur-Yvette.

I have benefited from many discussions with Guy Buss, Qun Chen, Brian Clarke, Andreas Dress, Gerd Faltings, Dan Freed, Dimitrij Leites, Manfred Liebmann, Xianqing Li-Jost, Jan Louis, Stephan Luckhaus, Kishore Marathe, René Meyer, Olaf Müller, Christoph Sachse, Klaus Sibold, Peter Teichner, Jürgen Tolksdorf, Guofang Wang, Shing-Tung Yau, Eberhard Zeidler, Miaomiao Zhu, and Kang Zuo. Several detailed computations for supersymmetric action functionals were supplied by Qun Chen, Abhijit Gadde, and René Meyer. Guy Buss, Brian Clarke, Christoph Sachse, Jürgen Tolksdorf and Miaomiao Zhu provided very useful lists of corrections and suggestions for clarifications and modifications. Minjie Chen helped me with some tex aspects, and he and Pengcheng Zhao created the figures, and Antje Vandenberg provided general logistic support. All this help and support I gratefully acknowledge.

Contents

Chapter 1
Geometry

1.1 Riemannian and Lorentzian Manifolds

1.1.1 Differential Geometry

We collect here some basic facts and principles of differential geometry as the foundation for the sequel. For a more penetrating discussion and for the proofs of various results, we refer to [65]. Classical differential geometry as expressed through the tensor calculus is about coordinate representations of geometric objects and the transformations of those representations under coordinate changes. The geometric objects are invariantly defined, but their coordinate representations are not, and resolving this contradiction is the content of the tensor calculus.

We consider a d-dimensional differentiable manifold M (assumed to be connected, oriented, paracompact and Hausdorff) and start with some conventions:

1. Einstein summation convention

$$a^i b_i := \sum_{i=1}^{d} a^i b_i. \tag{1.1.1}$$

The content of this convention is that a summation sign is omitted when the same index occurs twice in a product, once as an upper and once as a lower index. This rule is not affected by the possible presence of other indices; for example,

$$\Lambda^i_j b^j = \sum_{j=1}^{d} \Lambda^i_j b^j. \tag{1.1.2}$$

The conventions about when to place an index in an upper or lower position will be given subsequently. One aspect of this, however, is:

2. When $G = (g_{ij})_{i,j}$ is a metric tensor (a notion to be explained below) with indices i, j, the inverse metric tensor is written as $G^{-1} = (g^{ij})_{i,j}$, that is, by raising the indices. In particular

$$g^{ij} g_{jk} = \delta^i_k := \begin{cases} 1 & \text{when } i = k, \\ 0 & \text{when } i \neq k, \end{cases} \tag{1.1.3}$$

the so-called Kronecker symbol.

3. Combining the previous rules, we obtain more generally

$$v^i = g^{ij} v_j \quad \text{and} \quad v_i = g_{ij} v^j. \tag{1.1.4}$$

J. Jost, *Geometry and Physics*,
DOI 10.1007/978-3-642-00541-1_1, © Springer-Verlag Berlin Heidelberg 2009

4. For d-dimensional scalar quantities $(\phi^1, \ldots, \phi^d)$, we can use the Euclidean metric δ_{ij} to freely raise or lower indices in order to conform to the summation convention, that is,

$$\phi_i = \delta_{ij}\phi^j = \phi^i. \qquad (1.1.5)$$

A (finite-dimensional) manifold M is locally modeled after $\mathbb{R}^d$. Thus, locally, it can be represented by coordinates $x = (x^1, \ldots, x^d)$ taken from some open subset of $\mathbb{R}^d$. These coordinates, however, are not canonical, and we may as well choose other ones, $y = (y^1, \ldots, y^d)$, with $x = f(y)$ for some homeomorphism f. When the manifold M is differentiable—as always assumed here—we can cover it by local coordinates in such a manner that all such coordinate transitions are diffeomorphisms where defined. Again, the choice of coordinates is non-canonical. The basic content of classical differential geometry is to investigate how various expressions representing objects on M like tangent vectors transform under coordinate changes. Here and in the sequel, all objects defined on a differentiable manifold will be assumed to be differentiable themselves. This is checked in local coordinates, but since coordinate transitions are diffeomorphic, the differentiability property does not depend on the choice of coordinates.

Remark For our purposes, it is often convenient, and in the literature, it is customary, to mean by "differentiability" smoothness of class C^∞, that is, to assume that all objects are infinitely often differentiable. The ring of (infinitely often) differentiable functions on M is denoted by $C^\infty(M)$. Nonetheless, at certain places where analysis is more important, we need to be more specific about the regularity classes of the objects involved. But for the moment, we shall happily assume that our manifold M is of class C^∞.

A tangent vector for M at some point p represented by x_0 in local coordinates[1] x is an expression of the form

$$V = v^i \frac{\partial}{\partial x^i}. \qquad (1.1.6)$$

This means that it operates on a function $\phi(x)$ in our local coordinates as

$$V(\phi)(x_0) = v^i \frac{\partial \phi}{\partial x^i}\Big|_{x=x_0}. \qquad (1.1.7)$$

The summation convention (1.1.1) applies to (1.1.7). The i in $\frac{\partial}{\partial x^i}$ is considered to be a lower index since it appears in the denominator.

The tangent vectors at $p \in M$ form a vector space, called the tangent space T_pM of M at p. A basis of T_pM is given by the $\frac{\partial}{\partial x^i}$, considered as derivative operators

---

[1] We shall not always be so careful in distinguishing a point p as an invariant geometric object from its representation x_0 in some local coordinates, but frequently identify p and x_0 without alerting the reader.

at the point p represented by x_0 in the local coordinates, as in (1.1.7).[2] Whereas, as should become clear subsequently, this tangent space and its tangent vectors are defined independently of the choice of local coordinates, the representation of a tangent space does depend on those coordinates. The question then is how the same tangent vector is represented in different local coordinates y with $x = f(y)$ as before. The answer comes from the requirement that the result of the operation of the tangent vector V on a function ϕ, $V(\phi)$, be independent of the choice of coordinates. Always applying the chain rule, here and in the sequel, this yields

$$V = v^i \frac{\partial y^k}{\partial x^i} \frac{\partial}{\partial y^k}. \tag{1.1.8}$$

Thus, the coefficients of V in the y-coordinates are $v^i \frac{\partial y^k}{\partial x^i}$. This is verified by the following computation:

$$v^i \frac{\partial y^k}{\partial x^i} \frac{\partial}{\partial y^k} \phi(f(y)) = v^i \frac{\partial y^k}{\partial x^i} \frac{\partial \phi}{\partial x^j} \frac{\partial x^j}{\partial y^k} = v^i \frac{\partial x^j}{\partial x^i} \frac{\partial \phi}{\partial x^j} = v^i \frac{\partial \phi}{\partial x^i} \tag{1.1.9}$$

as required.

More abstractly, changing coordinates by f pulls a function ϕ defined in the x-coordinates back to $f^\star \phi$ defined for the y-coordinates, with $f^\star \phi(y) = \phi(f(y))$. If then $W = w^k \frac{\partial}{\partial y^k}$ is a tangent vector written in the y-coordinates, we need to push it forward as

$$f_\star W = w^k \frac{\partial x^i}{\partial y^k} \frac{\partial}{\partial x^i} \tag{1.1.10}$$

to the x-coordinates, to have the invariance

$$(f_\star W)(\phi) = W(f^\star \phi) \tag{1.1.11}$$

which is easily checked:

$$(f_\star W)\phi = w^k \frac{\partial x^i}{\partial y^k} \frac{\partial \phi}{\partial x^i} = w^k \frac{\partial}{\partial y^k} \phi(f(y)) = W(f^\star \phi). \tag{1.1.12}$$

In particular, there is some duality between functions and tangent vectors here. However, the situation is not entirely symmetric. We need to know the tangent vector only at the point x_0 where we want to apply it, but we need to know the function ϕ in some neighborhood of x_0 because we take its derivatives.

A vector field is then defined as $V(x) = v^i(x)\frac{\partial}{\partial x^i}$, that is, by having a tangent vector at each point of M. As indicated above, we assume here that the coefficients $v^i(x)$ are differentiable. The vector space of vector fields on M is written as $\Gamma(TM)$. (In fact, $\Gamma(TM)$ is a module over the ring $C^\infty(M)$.)

---

[2] As here, we shall usually simply write $\frac{\partial}{\partial x^i}$ in place of $\frac{\partial}{\partial x^i}(p)$ or $\frac{\partial}{\partial x^i}(x_0)$, that is, we assume that the point where a derivative operator acts is clear from the context or the coefficient.

Later, we shall need the Lie bracket $[V, W] := VW - WV$ of two vector fields $V(x) = v^i(x)\frac{\partial}{\partial x^i}$, $W(x) = w^j(x)\frac{\partial}{\partial x^j}$; its operation on a function ϕ is

$$[V, W]\phi(x) = v^i(x)\frac{\partial}{\partial x^i}\left(w^j(x)\frac{\partial}{\partial x^j}\phi(x)\right) - w^j(x)\frac{\partial}{\partial x^j}\left(v^i(x)\frac{\partial}{\partial x^i}\phi(x)\right)$$

$$= \left(v^i(x)\frac{\partial w^j(x)}{\partial x^i} - w^i(x)\frac{\partial v^j(x)}{\partial x^i}\right)\frac{\partial \phi(x)}{\partial x^j}. \qquad (1.1.13)$$

In particular, for coordinate vector fields, we have

$$\left[\frac{\partial}{\partial x^i}, \frac{\partial}{\partial x^j}\right] = 0. \qquad (1.1.14)$$

Returning to a single tangent vector, $V = v^i\frac{\partial}{\partial x^i}$ at some point x_0, we consider a covector or cotangent vector $\omega = \omega_i dx^i$ at this point as an object dual to V, with the rule

$$dx^i\left(\frac{\partial}{\partial x^j}\right) = \delta^i_j \qquad (1.1.15)$$

yielding

$$\omega_i dx^i\left(v^j\frac{\partial}{\partial x^j}\right) = \omega_i v^j \delta^i_j = \omega_i v^i. \qquad (1.1.16)$$

This expression depends only on the coefficients v^i and ω_i at the point under consideration and does not require any values in a neighborhood. We can write this as $\omega(V)$, the application of the covector ω to the vector V, or as $V(\omega)$, the application of V to ω.

The cotangent vectors at p likewise constitute a vector space, the cotangent space $T_p^\star M$.

We have the transformation behavior

$$dx^i = \frac{\partial x^i}{\partial y_\alpha}dy^\alpha \qquad (1.1.17)$$

required for the invariance of $\omega(V)$. Thus, the coefficients of ω in the y-coordinates are given by the identity

$$\omega_i dx^i = \omega_i \frac{\partial x^i}{\partial y_\alpha}dy^\alpha. \qquad (1.1.18)$$

Again, a covector $\omega_i dx^i$ is pulled back under a map f:

$$f^\star(\omega_i dx^i) = \omega_i \frac{\partial x^i}{\partial y^\alpha}dy^\alpha. \qquad (1.1.19)$$

The transformation rules (1.1.10), (1.1.19) apply to arbitrary maps $f : M \to N$ from M into a possibly different manifold N, not only to coordinate changes or diffeomorphisms. So, we can always pull back a function or a covector and always push forward a vector under a map, but not always the other way around.

The transformation behavior of a tangent vector as in (1.1.8) is called contravariant, the opposite one of a covector as (1.1.18) covariant.

A 1-form then assigns a covector to every point in M, and thus, it is locally given as $\omega_i(x)dx^i$.

Having derived the transformation of vectors and covectors, we can then also determine the transformation rules for other tensors. **A lower index always indicates covariant, an upper one contravariant transformation.** For example, the metric tensor, written as $g_{ij}dx^i \otimes dx^j$,[3] with $g_{ij} = \langle \frac{\partial}{\partial x^i}, \frac{\partial}{\partial x^j} \rangle$ being the inner product of those two basis vectors, operates on pairs of tangent vectors. It therefore transforms doubly covariantly, that is, becomes

$$g_{ij}(f(y))\frac{\partial x^i}{\partial y^\alpha}\frac{\partial x^j}{\partial y^\beta}dy^\alpha \otimes dy^\beta. \tag{1.1.20}$$

The purpose of the metric tensor is to provide a Euclidean product of tangent vectors,

$$\langle V, W \rangle = g_{ij}v^i w^j \tag{1.1.21}$$

for $V = v^i \frac{\partial}{\partial x^i}$, $W = w^i \frac{\partial}{\partial x^i}$. As a check, in this formula, v^i and w^i transform contravariantly, while g_{ij} transforms doubly covariantly, so that the product as a scalar quantity remains invariant under coordinate transformations.

Similarly, we obtain the product of two covectors $\omega, \alpha \in T_x^\star M$ as

$$\langle \omega, \alpha \rangle = g^{ij}\omega_i\alpha_j. \tag{1.1.22}$$

We next introduce the concept of exterior p-forms and put

$$\Lambda^p := \Lambda^p(T_x^\star M) := \underbrace{T_x^\star M \wedge \cdots \wedge T_x^\star M}_{p \text{ times}} \quad \text{(exterior product)}. \tag{1.1.23}$$

On $\Lambda^p(T_x^\star M)$, we have the exterior product with $\eta \in T_x^\star M = \Lambda^1(T_x^\star M)$:

$$\Lambda^p(T_x^\star M) \longrightarrow \Lambda^{p+1}(T_x^\star M)$$
$$\omega \longmapsto \epsilon(\eta)\omega := \eta \wedge \omega. \tag{1.1.24}$$

An exterior p-form is a sum of terms of the form

$$\omega(x) = \eta(x)dx^{i_1} \wedge \cdots \wedge dx^{i_p}$$

---

[3] Subsequently, we shall mostly leave out the symbol $\otimes$, that is, write simply $g_{ij}dx^i dx^j$ in place of $g_{ij}dx^i \otimes dx^j$.

where $\eta(x)$ is a smooth function and $(x^1, \ldots, x^d)$ are local coordinates. That is, a p-form assigns an element of $\Lambda^p(T_x^\star M)$ to every $x \in M$. The space of exterior p-forms is denoted by $\Omega^p(M)$.

When M carries a Riemannian metric $g_{ij}dx^i \otimes dx^j$, the scalar product on the cotangent spaces $T_x^\star M$ induces one on the spaces $\Lambda^p(T_x^\star M)$ by

$$\langle dx^{i_1} \wedge \cdots \wedge dx^{i_p}, dx^{j_1} \wedge \cdots \wedge dx^{j_p}\rangle := \det((\langle dx^{i_\mu}, dx^{j_\nu}\rangle)) \qquad (1.1.25)$$

and linear extension.

Given a Riemannian metric $g_{ij}dx^i \otimes dx^j$, also, in local coordinates, we can define the volume form

$$dvol_g := \sqrt{\det(g_{ij})}dx^1 \wedge \cdots \wedge dx^d. \qquad (1.1.26)$$

This volume form depends on an ordering of the indices $1, 2, \ldots, d$ of the local coordinates: since the exterior product is antisymmetric, $dx^i \wedge dx^j = -dx^j \wedge dx^i$, it changes its sign under an odd permutation of the indices. Thus, when we have a coordinate transformation $x = f(y)$ where the Jacobian determinant $\det(\frac{\partial x^i}{\partial y^\alpha})$ is negative, $dvol$ changes its sign; otherwise, it is invariant. Therefore, in order to have a globally defined volume form on the Riemannian manifold M, we need to exclude coordinate changes with negative Jacobian. The manifold M is called *oriented* when it can be covered by coordinates such that all coordinate changes have a positive Jacobian. In that case, the volume form is well defined, and we can define the integral of a function ϕ on M by

$$\int \phi(x)\, dvol_g(x). \qquad (1.1.27)$$

We shall therefore assume the manifold M to be oriented whenever we carry out such an integral. We can then also define the L^2-product of p-forms $\omega, \alpha \in \Omega^p(M)$:

$$(\omega, \alpha) := \int \langle \omega(x), \alpha(x)\rangle dvol_g(x). \qquad (1.1.28)$$

We now assume that the dimension $d = 4$, the case of particular importance for the application of our geometric concepts to physics. Then when ω is a 2-form, $\omega \wedge \omega$ is a 4-form. We call ω self-dual or antiself-dual when the $+$ resp. $-$ sign holds in

$$\omega \wedge \omega = \pm\langle \omega, \omega\rangle dvol_g. \qquad (1.1.29)$$

When ω_+ is self-dual, and ω_- antiself-dual, we have

$$\langle \omega_+, \omega_-\rangle = 0 \qquad (1.1.30)$$

that is, the spaces of self-dual and antiself-dual forms are orthogonal to each other. Every 2-form ω on a 4-manifold can be decomposed as the sum of a self-dual and an antiself-dual form,

$$\omega = \omega_+ + \omega_-. \qquad (1.1.31)$$

We return to arbitrary dimension d.

Definition 1.1 The *exterior derivative* $d : \Omega^p(M) \to \Omega^{p+1}(M)$ $(p = 0, \ldots, \dim M)$ is defined through the formula

$$d(\eta(x)dx^{i_1} \wedge \cdots \wedge dx^{i_p}) = \frac{\partial \eta(x)}{\partial x^j}dx^j \wedge dx^{i_1} \wedge \cdots \wedge dx^{i_p} \qquad (1.1.32)$$

and extended by linearity to all of $\Omega^p(M)$.

The exterior derivative enjoys the following product rule: If $\omega \in \Omega^p(M)$, $\vartheta \in \Omega^q(M)$, then

$$d(\omega \wedge \vartheta) = d\omega \wedge \vartheta + (-1)^p \omega \wedge d\vartheta, \qquad (1.1.33)$$

from the formula $\omega \wedge \vartheta = (-1)^{pq}\vartheta \wedge \omega$ and (1.1.32).

Let $x = f(y)$ be a coordinate transformation,

$$\omega(x) = \eta(x)dx^{i_1} \wedge \cdots \wedge dx^{i_p} \in \Omega^p(M).$$

In the y-coordinates, we then have

$$f^*(\omega)(y) = \eta(f(y))\frac{\partial x^{i_1}}{\partial y^{\alpha_1}}dy^{\alpha_1} \wedge \cdots \wedge \frac{\partial x^{i_p}}{\partial y^{\alpha_p}}dy^{\alpha_p} \qquad (1.1.34)$$

which is the transformation formula for p-forms. The exterior derivative is compatible with this transformation rule:

$$d(f^*(\omega)) = f^*(d\omega), \qquad (1.1.35)$$

which follows from the transformation invariance

$$\frac{\partial \eta(x)}{\partial x^j}dx^j = \frac{\partial \eta(f(y))}{\partial x^j}\frac{\partial f^j}{\partial y^\alpha}dy^\alpha = \frac{\partial \eta(f(y))}{\partial y^\alpha}dy^\alpha. \qquad (1.1.36)$$

Thus, d is independent of the choice of coordinates. d satisfies the following important rule:

Lemma 1.1

$$d \circ d = 0. \qquad (1.1.37)$$

Proof We check (1.1.37) for forms of the type

$$\omega(x) = f(x)dx^{i_1} \wedge \cdots \wedge dx^{i_p}$$

from which it extends by linearity to all p-forms. Now

$$d \circ d(\omega(x)) = d\left(\frac{\partial f}{\partial x^j}dx^j \wedge dx^{i_1} \wedge \cdots \wedge dx^{i_p}\right)$$

$$= \frac{\partial^2 f}{\partial x^j \partial x^k}dx^k \wedge dx^j \wedge dx^{i_1} \wedge \cdots \wedge dx^{i_p} = 0,$$

since $\frac{\partial^2 f}{\partial x^j \partial x^k} = \frac{\partial^2 f}{\partial x^k \partial x^j}$ and $dx^j \wedge dx^k = -dx^k \wedge dx^j$. □

In the preceding, we have presented one possible way of conceptualizing trans-formations, the one employed by mathematicians: The same point p is written in different coordinate systems x and y, which are then functionally related by $x = x(y)$. Another view of transformations, often taken in the physics literature, is to move the point p and consider the induced effect on tensors. Let us discuss the example of a 1-form $\omega(x)dx$. Within the fixed coordinates x, we vary the points represented by these coordinates by

$$x \mapsto x + \epsilon \xi(x) =: x + \epsilon \delta x \tag{1.1.38}$$

for some map ξ and some small parameter ϵ, and we want to take the limit $\epsilon \to 0$. We have the induced variation of our 1-form

$$\omega(x)dx \mapsto \omega(x + \epsilon \xi(x))d(x + \epsilon \xi(x)) =: \omega(x) + \epsilon \delta \omega(x). \tag{1.1.39}$$

By Taylor expansion, we have

$$\omega(x + \epsilon \xi(x))d(x + \epsilon \xi(x)) = \left(\omega_i(x) + \epsilon \frac{\partial \omega_i}{\partial x^k}\xi^k(x)\right)\left(dx^i + \epsilon \frac{\partial \xi^i}{\partial x^k}dx^k\right)$$

$$+ \text{ higher order terms} \tag{1.1.40}$$

from which we conclude that for $\epsilon \to 0$

$$\delta \omega = \frac{\partial \omega_i}{\partial x^k}\xi^k dx^i + \omega_i \frac{\partial \xi^i}{\partial x^k}dx^k. \tag{1.1.41}$$

Of course, since $\frac{\partial \xi^i}{\partial x^k}dx^k = d\xi^i$, the last term in (1.1.41) agrees with the one required by (1.1.18).

To put the preceding into a slogan: *For setting up transformation rules in geometry, mathematicians keep the point fixed and change the coordinates, while physicists keep the same coordinates, but move the point around.* The first approach is well suited to identifying invariants, like the curvature tensor. The second one is convenient for computing variations, as in our discussion of actions below.

So far, we have computed derivatives of functions. We have also talked about vector fields $V(x) = v^i(x)\frac{\partial}{\partial x^i}$ as objects that depend differentiably on their arguments x. Of course, we can do the same for other tensors, like the metric $g_{ij}(x)dx^i \otimes dx^j$. This naturally raises the question about how to compute their derivatives. This encounters the problem, however, that in contrast to functions, the representation of such tensors depends on the choice of local coordinates, and we have described in some detail that and how they transform under coordinate changes. Precisely because of that transformation, they acquire a coordinate invariant meaning; for example, the operation of a vector on a function or the metric product between two vectors are both independent of the choice of coordinates.

It now turns out that on a differentiable manifold, there is in general no single canonical way of taking derivatives of vector fields or other tensors in an invariant manner. There are, in fact, many such possibilities, and they are called connections or covariant derivatives. Only when we have additional structures, like a Riemannian metric, can we single out a particular covariant derivative on the basis of its compatibility with the metric. For our purposes, however, we also need other covariant derivatives, and therefore, we now develop that notion. We shall treat this issue from a more abstract perspective in Sect. 1.2 below, and so the reader who wants to progress more rapidly can skip the discussion here.

Let M be a differentiable manifold. We recall that $\Gamma(TM)$ denotes the space of vector fields on M. An (affine) connection or covariant derivative on M is a linear map

$$\nabla : \Gamma(TM) \otimes_{\mathbb{R}} \Gamma(TM) \to \Gamma(TM),$$
$$(V, W) \mapsto \nabla_V W$$

satisfying:

(i) ∇ is tensorial in the first argument:

$$\nabla_{V_1+V_2} W = \nabla_{V_1} W + \nabla_{V_2} W \quad \text{for all } V_1, V_2, W \in \Gamma(TM),$$
$$\nabla_{fV} W = f \nabla_V W \quad \text{for all } f \in C^\infty(M), V, W \in \Gamma(TM);$$

(ii) ∇ is $\mathbb{R}$-linear in the second argument:

$$\nabla_V (W_1 + W_2) = \nabla_V W_1 + \nabla_V W_2 \quad \text{for all } V, W_1, W_2 \in \Gamma(TM)$$

and it satisfies the product rule

$$\nabla_V (fW) = V(f)W + f\nabla_V W \quad \text{for all } f \in C^\infty(M), V, W \in \Gamma(TM). \tag{1.1.42}$$

$\nabla_V W$ is called the covariant derivative of W in the direction V. By (i), for any $x_0 \in M$, $(\nabla_V W)(x_0)$ only depends on the value of V at x_0. By way of contrast, it also depends on the values of W in some neighborhood of x_0, as it naturally should as a notion of a derivative of W. The example on which this is modeled is the Euclidean connection given by the standard derivatives, that is, for $V = V^i \frac{\partial}{\partial x^i}$, $W = W^j \frac{\partial}{\partial x^j}$,

$$\nabla_V^{eucl} W = V^i \frac{\partial W^j}{\partial x^i} \frac{\partial}{\partial x^j}.$$

However, this is not invariant under nonlinear coordinate changes, and since a general manifold cannot be covered by coordinates with only linear coordinate transformations, we need the above more general and abstract concept of a covariant derivative.

Let U be a coordinate chart in M, with local coordinates x and coordinate vector fields $\frac{\partial}{\partial x^1}, \ldots, \frac{\partial}{\partial x^d}$ ($d = \dim M$). We then define the Christoffel symbols of the connection ∇ via

$$\nabla_{\frac{\partial}{\partial x^i}} \frac{\partial}{\partial x^j} =: \Gamma_{ij}^k \frac{\partial}{\partial x^k}. \tag{1.1.43}$$

Thus,

$$\nabla_V W = V^i \frac{\partial W^j}{\partial x^i} \frac{\partial}{\partial x^j} + V^i W^j \Gamma_{ij}^k \frac{\partial}{\partial x^k}. \tag{1.1.44}$$

In order to understand the nature of the objects involved, we can also leave out the vector field V and consider the covariant derivative ∇W as a 1-form. In local coordinates

$$\nabla W = W_{;i}^j \frac{\partial}{\partial x^j} dx^i, \tag{1.1.45}$$

with

$$W_{;i}^j := \frac{\partial W^j}{\partial x^i} + W^k \Gamma_{ik}^j. \tag{1.1.46}$$

If we change our coordinates x to coordinates y, then the new Christoffel symbols,

$$\nabla_{\frac{\partial}{\partial y^l}} \frac{\partial}{\partial y^m} =: \tilde{\Gamma}_{lm}^n \frac{\partial}{\partial y^n}, \tag{1.1.47}$$

are related to the old ones via

$$\tilde{\Gamma}_{lm}^n(y(x)) = \left\{ \Gamma_{ij}^k(x) \frac{\partial x^i}{\partial y^l} \frac{\partial x^j}{\partial y^m} + \frac{\partial^2 x^k}{\partial y^l \partial y^m} \right\} \frac{\partial y^n}{\partial x^k}. \tag{1.1.48}$$

In particular, due to the term $\frac{\partial^2 x^k}{\partial y^l \partial y^m}$, the Christoffel symbols do not transform as a tensor. However, if we have two connections $^1\nabla, {}^2\nabla$, with corresponding Christoffel symbols $^1\Gamma_{ij}^k, {}^2\Gamma_{ij}^k$, then the difference $^1\Gamma_{ij}^k - {}^2\Gamma_{ij}^k$ does transform as a tensor. Expressed more abstractly, this means that the space of connections on M is an affine space.

For a connection ∇, we define its torsion tensor via

$$T(V, W) := \nabla_V W - \nabla_W V - [V, W] \quad \text{for } V, W \in \Gamma(TM). \tag{1.1.49}$$

Inserting our coordinate vector fields $\frac{\partial}{\partial x^i}$ as before, we obtain

$$T_{ij} := T\left(\frac{\partial}{\partial x^i}, \frac{\partial}{\partial x^j} \right) = \nabla_{\frac{\partial}{\partial x^i}} \frac{\partial}{\partial x^j} - \nabla_{\frac{\partial}{\partial x^j}} \frac{\partial}{\partial x^i}$$

(since coordinate vector fields commute, i.e., $[\frac{\partial}{\partial x^i}, \frac{\partial}{\partial x^j}] = 0$)

$$(\Gamma^k_{ij} - \Gamma^k_{ji})\frac{\partial}{\partial x^k}.$$

We call the connection ∇ torsion-free or symmetric if $T \equiv 0$. By the preceding computation, this is equivalent to the symmetry

$$\Gamma^k_{ij} = \Gamma^k_{ji} \quad \text{for all } i, j, k. \tag{1.1.50}$$

Let $c(t)$ be a smooth curve in M, and let $V(t) := \dot{c}(t) \; (= \dot{c}^i(t)\frac{\partial}{\partial x^i}(c(t))$ in local coordinates) be the tangent vector field of c. In fact, we should instead write $V(c(t))$ in place of $V(t)$, but we consider t as the coordinate along the curve $c(t)$. Thus, in those coordinates $\frac{\partial}{\partial t} = \frac{\partial c^i}{\partial t}\frac{\partial}{\partial x^i}$, and in the sequel, we shall frequently and implicitly make this identification, that is, switch between the points $c(t)$ on the curve and the corresponding parameter values t. Let $W(t)$ be another vector field along c, i.e., $W(t) \in T_{c(t)}M$ for all t. We may then write $W(t) = \mu^i(t)\frac{\partial}{\partial x^i}(c(t))$ and form

$$\nabla_{\dot{c}(t)} W(t) = \dot{\mu}^i(t)\frac{\partial}{\partial x^i} + \dot{c}^i(t)\mu^j(t)\nabla_{\frac{\partial}{\partial x^i}}\frac{\partial}{\partial x^j}$$

$$= \dot{\mu}^i(t)\frac{\partial}{\partial x^i} + \dot{c}^i(t)\mu^j(t)\Gamma^k_{ij}(c(t))\frac{\partial}{\partial x^k}$$

(the preceding computation is meaningful as we see that it depends only on the values of W along the curve $c(t)$, but not on other values in a neighborhood of a point on that curve).

This represents a (nondegenerate) linear system of d first-order differential operators for the d coefficients $\mu^i(t)$ of $W(t)$. Therefore, for given initial values $\mu^i(0)$, there exists a unique solution $W(t)$ of

$$\nabla_{\dot{c}(t)} W(t) = 0.$$

This $W(t)$ is called the parallel transport of $W(0)$ along the curve $c(t)$. We also say that $W(t)$ is covariantly constant along the curve c.

Now, let W be a vector field in a neighborhood U of some point $x_0 \in M$. W is called parallel if for any curve $c(t)$ in U, $W(t) := W(c(t))$ is parallel along c. This means that for all tangent vectors V in U,

$$\nabla_V W = 0,$$

i.e.,

$$\frac{\partial}{\partial x^i}W^k + W^j\Gamma^k_{ij} = 0 \quad \text{identically in } U, \text{ for all } i, k,$$

$$\text{with } W = W^i\frac{\partial}{\partial x^i} \text{ in local coordinates.}$$

This now is a system of d^2 first-order differential equations for the d coefficients of W, and so, it is overdetermined. Therefore, in general, such W do not exist. Of course, they do exist for the Euclidean connection, because in Euclidean coordinates, the coordinate vector fields $\frac{\partial}{\partial x^i}$ are parallel.

We define the curvature tensor R by

$$R(V, W)Z := \nabla_V \nabla_W Z - \nabla_W \nabla_V Z - \nabla_{[V,W]}Z, \tag{1.1.51}$$

or in local coordinates

$$R^k_{lij} \frac{\partial}{\partial x^k} := R\left(\frac{\partial}{\partial x^i}, \frac{\partial}{\partial x^j}\right) \frac{\partial}{\partial x^l} \quad (i, j, l = 1, \ldots, d). \tag{1.1.52}$$

The curvature tensor can be expressed in terms of the Christoffel symbols and their derivatives via

$$R^k_{lij} = \frac{\partial}{\partial x^i} \Gamma^k_{jl} - \frac{\partial}{\partial x^j} \Gamma^k_{il} + \Gamma^k_{im} \Gamma^m_{jl} - \Gamma^k_{jm} \Gamma^m_{il}. \tag{1.1.53}$$

We also note that, as the name indicates, the curvature tensor R is, like the torsion tensor T, but in contrast to the connection ∇ represented by the Christoffel symbols, a tensor. This means that when one of its arguments is multiplied by a smooth function, we may simply pull out that function without having to take a derivative of it. Equivalently, it transforms as a tensor under coordinate changes; here, the upper index k stands for an argument that transforms as a vector, that is contravariantly, whereas the lower indices l, i, j express a covariant transformation behavior. The curvature tensor will be discussed in more detail in Sect. 1.1.5.

A curve $c(t)$ in M is called autoparallel or geodesic if

$$\nabla_{\dot{c}} \dot{c} = 0. \tag{1.1.54}$$

Geodesics will be discussed in detail and from a different perspective in Sect. 1.1.4. Here, we only display their equation and define the exponential map. In local coordinates, (1.1.54) becomes

$$\ddot{c}^k(t) + \Gamma^k_{ij}(c(t))\dot{c}^i(t)\dot{c}^j(t) = 0 \quad \text{for } k = 1, \ldots, d. \tag{1.1.55}$$

This constitutes a system of second-order ODEs, and given $x_0 \in M$, $V \in T_{x_0}M$, there exist a maximal interval $I_V \subset \mathbb{R}$ containing an open neighborhood of 0 and a geodesic

$$c_V : I_V \to M$$

with $c_V(0) = x_0$, $\dot{c}_V(0) = V$. We can then define the exponential map $\exp_{x_0}$ on some star-shaped neighborhood of $0 \in T_{x_0}M$:

$$\exp_{x_0} : \{V \in T_{x_0}M : 1 \in I_V\} \to M,$$
$$V \mapsto c_V(1). \tag{1.1.56}$$

We then have $\exp_{x_0}(tV) = c_V(t)$ for $0 \le t \le 1$.

A submanifold S of M is called autoparallel or totally geodesic if for all $x_0 \in S$, $V \in T_{x_0} S$ for which $\exp_{x_0} V$ is defined, we have

$$\exp_{x_0} V \in S.$$

The infinitesimal condition needed for this property is that

$$\nabla_V W(x) \in T_x S$$

for any vector field $W(x)$ tangent to S and $V \in T_x S$.

Now, let M carry a Riemannian metric $g = \langle \cdot, \cdot \rangle$.

We say that ∇ is a Riemannian connection if it satisfies the metric product rule

$$Z\langle V, W \rangle = \langle \nabla_Z V, W \rangle + \langle V, \nabla_Z W \rangle. \tag{1.1.57}$$

For any Riemannian metric g, there exists a unique torsion-free Riemannian connection, the so-called Levi-Città connection ∇^g. It is given by

$$\langle \nabla^g_V W, Z \rangle = \frac{1}{2}\{V\langle W, Z \rangle - Z\langle V, W \rangle + W\langle Z, V \rangle$$
$$- \langle V, [W, Z] \rangle + \langle Z, [V, W] \rangle + \langle W, [Z, V] \rangle\}. \tag{1.1.58}$$

The Christoffel symbols of ∇^g can be expressed through the metric; in local coordinates, with $g_{ij} = \langle \frac{\partial}{\partial x^i} \frac{\partial}{\partial x^j} \rangle$, we use the abbreviation

$$g_{ij,k} := \frac{\partial}{\partial x^k} g_{ij} \tag{1.1.59}$$

and have

$$\Gamma^k_{ij} = \frac{1}{2} g^{kl} (g_{il,j} + g_{jl,i} - g_{ij,l}), \tag{1.1.60}$$

or, equivalently,

$$g_{ij,k} = g_{jl}\Gamma^l_{ik} + g_{il}\Gamma^l_{jk} = \Gamma_{ikj} + \Gamma_{jki}. \tag{1.1.61}$$

The Levi-Città connection ∇^g respects the metric in the sense that if $V(t), W(t)$ are parallel vector fields along a curve $c(t)$, then

$$\langle V(t), W(t) \rangle \equiv \text{const}, \tag{1.1.62}$$

that is, products between tangent vectors remain invariant under parallel transport.

1.1.2 Complex Manifolds

We start with complex dimension 1. The Euclidean space $\mathbb{R}^2$ can be made into the complex vector space $\mathbb{C}^1$ on which multiplication by complex numbers of the form $a + ib$ is defined, with $i = \sqrt{-1}$. Conventions:

$$z = x + iy = x^1 + ix^2, \quad \bar{z} = x - iy. \tag{1.1.63}$$

In the physics literature, z and $\bar{z}$ are formally viewed as independent coordinates. We define

$$\partial_z := \frac{\partial}{\partial z} = \frac{1}{2}(\partial_x - i\partial_y), \qquad \partial_{\bar{z}} = \frac{\partial}{\partial \bar{z}} = \frac{1}{2}(\partial_x + i\partial_y). \tag{1.1.64}$$

This is arranged so that

$$\partial_z z = 1, \qquad \partial_z \bar{z} = 0, \tag{1.1.65}$$

and so on. A function $f : \mathbb{C} \to \mathbb{C}$ is called holomorphic if

$$\partial_{\bar{z}} f = 0. \tag{1.1.66}$$

Mathematicians write $f(z)$ for any function of the complex variable z. Physicists instead write $f(z, \bar{z})$, reserving the notation $f(z)$ for a holomorphic function, that is, one satisfying (1.1.66) because that relation formally expresses independence of the coordinate $\bar{z}$. Similarly, $g : \mathbb{C} \to \mathbb{C}$ is antiholomorphic if

$$\partial_z g = 0. \tag{1.1.67}$$

Another reason for the physics convention is to consider the complexification $\mathbb{C}^2$ with coordinates (z, z') of the Euclidean plane $\mathbb{C} = \mathbb{R}^2$. The slice defined by $\bar{z} = z'$ then yields the Euclidean plane, while $(z, z') = i(s + t, s - t)$ gives the Minkowski plane with metric $dt^2 - ds^2$.

When we use the conformal transformation $z = e^w$, with $w = \tau + i\sigma$, $-\infty < \tau < \infty$ and $0 \le \sigma < 2\pi$, and pass from $w = \tau + i\sigma$ to the light cone coordinates $\zeta^+ = \tau + \sigma$, $\zeta^- = \tau - \sigma$ (a so-called Wick rotation), we obtain the Minkowski metric in the form $d\zeta^+ d\zeta^-$.

In complex coordinates, the Laplace operator (see (1.1.103), (1.1.105) below) becomes

$$\Delta = \frac{\partial^2}{\partial x^2} + \frac{\partial^2}{\partial y^2} = 4\frac{\partial^2}{\partial z \partial \bar{z}}. \tag{1.1.68}$$

We next have the 1-forms

$$dz = dx + i\,dy, \qquad d\bar{z} = dx - i\,dy. \tag{1.1.69}$$

This is arranged so that

$$dz(\partial_z) = 1, \qquad dz(\partial_{\bar{z}}) = 0, \tag{1.1.70}$$

and so on, the analogs of (1.1.15). For a vector $v^1 \frac{\partial}{\partial x} + v^2 \frac{\partial}{\partial y}$, we write

$$v^z := v^1 + iv^2, \qquad v^{\bar{z}} := v^1 - iv^2, \tag{1.1.71}$$

and (in flat space)

$$v_z := \frac{1}{2}(v^1 - iv^2), \qquad v_{\bar{z}} := \frac{1}{2}(v^1 + iv^2). \tag{1.1.72}$$

In this notation, the Euclidean (flat) metric on $\mathbb{R}^2$, $g_{11} = g_{22} = 1$, $g_{12} = 0$, becomes

$$g_{z\bar{z}} = g_{\bar{z}z} = \frac{1}{2}, \qquad g_{zz} = g_{\bar{z}\bar{z}} = 0, \qquad g^{z\bar{z}} = g^{\bar{z}z} = 2, \qquad g^{zz} = g^{\bar{z}\bar{z}} = 0. \tag{1.1.73}$$

This is set up to be compatible with (1.1.4). Thus, (1.1.72) becomes a special case of

$$v_z = g_{zz} v^z + g_{z\bar{z}} v^{\bar{z}}. \tag{1.1.74}$$

The area form for this metric is

$$\frac{i}{2} dz \wedge d\bar{z} = dx \wedge dy. \tag{1.1.75}$$

The conventions become clearer when we observe

$$\sqrt{g_{11} g_{22} - g_{12}^2}\, dx \wedge dy = \sqrt{g_{zz} g_{\bar{z}\bar{z}} - g_{z\bar{z}}^2}\, dz \wedge d\bar{z}. \tag{1.1.76}$$

Also, for a twice covariant tensor,

$$V_{zz} = \frac{1}{4}(V_{11} + 2i V_{12} - V_{22}), \qquad V_{\bar{z}\bar{z}} = \frac{1}{4}(V_{11} - 2i V_{12} - V_{22}),$$

$$V_{z\bar{z}} = V_{\bar{z}z} = \frac{1}{4}(V_{11} + V_{22}) \tag{1.1.77}$$

of which (1.1.73) is a special case.

The divergence is (in flat space)

$$\partial_{x^1} v^1 + \partial_{x^2} v^2 = \partial_z v^z + \partial_{\bar{z}} v^{\bar{z}}. \tag{1.1.78}$$

The divergence theorem (integration by parts, a special case of Stokes' theorem) is here

$$\int_\Omega (\partial_z v^z + \partial_{\bar{z}} v^{\bar{z}}) \frac{i}{2} dz \wedge d\bar{z} = \frac{i}{2} \oint_{\partial\Omega} (v^z d\bar{z} - v^{\bar{z}} dz) \tag{1.1.79}$$

with a counterclockwise contour integral around Ω.

We now turn to the higher-dimensional situation. The model space is now $\mathbb{C}^d$, the d-dimensional complex vector space. The preceding expressions defined for $d = 1$ then get equipped with coordinate indices:

$$z = (z^1, \ldots, z^d), \quad \text{with } z^j = x^j + i y^j \tag{1.1.80}$$

using $(x^1, y^1, \ldots, x^d, y^d)$ as Euclidean coordinates on $\mathbb{R}^{2d}$, and

$$z^{\bar{j}} := x^j - i y^j.$$

Likewise

$$\partial_{\bar{k}} := \frac{\partial}{\partial z^{\bar{k}}} := \frac{1}{2}\left(\frac{\partial}{\partial x^k} + i \frac{\partial}{\partial y^k}\right), \tag{1.1.81}$$

and so on. Then, a function $f : \mathbb{C}^d \to \mathbb{C}$ is holomorphic if

$$\partial_{\bar{k}} f = 0 \tag{1.1.82}$$

for $k = 1, \ldots, d$.

Definition 1.2 A *complex manifold* of complex dimension d ($\dim_{\mathbb{C}} M = d$) is a differentiable manifold of (real) dimension $2d$ ($\dim_{\mathbb{R}} M = 2d$) whose charts take values in open subsets of $\mathbb{C}^d$ with *holomorphic* coordinate transitions.

A one-dimensional complex manifold is also called a Riemann surface, but that subject will be taken up in more depth in Sect. 1.4.2 below.

Let M again be a complex manifold of complex dimension d. Let $T_z^{\mathbb{R}} M := T_z M$ be the ordinary (real) tangent space of M at z. We define the complexified tangent space

$$T_z^{\mathbb{C}} M := T_z^{\mathbb{R}} M \otimes_{\mathbb{R}} \mathbb{C} \tag{1.1.83}$$

which we then decompose as

$$T_z^{\mathbb{C}} M = \mathbb{C}\left\{ \frac{\partial}{\partial z^j}, \frac{\partial}{\partial \bar{z}^j} \right\} =: T_z' M \oplus T_z'' M, \tag{1.1.84}$$

where $T_z' M = \mathbb{C}\{\frac{\partial}{\partial z^j}\}$ is the holomorphic and $T_z'' M = \mathbb{C}\{\frac{\partial}{\partial \bar{z}^j}\}$ the antiholomorphic tangent space. In $T_z^{\mathbb{C}} M$, we have a conjugation mapping $\frac{\partial}{\partial z^j}$ to $\frac{\partial}{\partial \bar{z}^j}$, and so $T_z'' M = \overline{T_z' M}$. The same construction is possible for the cotangent space, and we have analogously

$$T_z^{\star \mathbb{C}} M = \mathbb{C}\{dz^j, d\bar{z}^j\} =: T_z^{\star'} M \oplus T_z^{\star''} M. \tag{1.1.85}$$

The important point is that these decompositions are invariant under coordinate changes because those coordinate changes are required to be holomorphic. In particular, we have the transformation rules

$$dz^j = \frac{\partial z^j}{\partial w^l} dw^l, \qquad d\bar{z}^{\bar{k}} = \left(\overline{\frac{\partial z^k}{\partial w^m}} \right) dw^{\bar{m}} = \frac{\partial \bar{z}^{\bar{k}}}{\partial w^{\bar{m}}} dw^{\bar{m}} \tag{1.1.86}$$

when $z = z(w)$.

The complexified space $\Omega^k(M; \mathbb{C})$ of k-forms can be decomposed into subspaces $\Omega^{p,q}(M)$ with $p + q = k$. $\Omega^{p,q}(M)$ is locally spanned by forms of the type

$$\omega(z) = \eta(z) dz^{i_1} \wedge \cdots \wedge dz^{i_p} \wedge d\bar{z}^{\bar{j}_1} \wedge \cdots \wedge d\bar{z}^{\bar{j}_q}. \tag{1.1.87}$$

Thus

$$\Omega^k(M) = \bigoplus_{p+q=k} \Omega^{p,q}(M). \tag{1.1.88}$$

We can then let the differential operators

$$\partial = \frac{1}{2}\left(\frac{\partial}{\partial x^j} - i\frac{\partial}{\partial y^j}\right)(dx^j + idy^j) \quad \text{and}$$

$$\bar{\partial} = \frac{1}{2}\left(\frac{\partial}{\partial x^j} + i\frac{\partial}{\partial y^j}\right)(dx^j - idy^j)$$

$(1.1.89)$

operate on such a form by

$$\partial\omega = \frac{\partial\eta}{\partial z^i}dz^i \wedge dz^{i_1} \wedge \cdots \wedge dz^{i_p} \wedge dz^{\bar{j}_1} \wedge \cdots \wedge dz^{\bar{j}_q} \qquad (1.1.90)$$

and

$$\bar{\partial}\omega = \frac{\partial\eta}{\partial z^{\bar{j}}}dz^{\bar{j}} \wedge dz^{i_1} \wedge \cdots \wedge dz^{i_p} \wedge dz^{\bar{j}_1} \wedge \cdots \wedge dz^{\bar{j}_q}. \qquad (1.1.91)$$

∂ and $\bar{\partial}$ yield a decomposition of the exterior derivative d:

Lemma 1.2

$$d = \partial + \bar{\partial}. \qquad (1.1.92)$$

Moreover,

$$\partial\partial = 0, \qquad \bar{\partial}\bar{\partial} = 0, \qquad (1.1.93)$$

$$\partial\bar{\partial} = -\bar{\partial}\partial. \qquad (1.1.94)$$

Proof

$$\partial + \bar{\partial} = \frac{1}{2}\left(\frac{\partial}{\partial x^j} - i\frac{\partial}{\partial y^j}\right)(dx^j + idy^j) + \frac{1}{2}\left(\frac{\partial}{\partial x^j} + i\frac{\partial}{\partial y^j}\right)(dx^j - idy^j)$$

$$= \frac{\partial}{\partial x^j}dx^j + \frac{\partial}{\partial y^j}dy^j = d.$$

Consequently,

$$0 = d^2 = (\partial + \bar{\partial})(\partial + \bar{\partial}) = \partial^2 + \partial\bar{\partial} + \bar{\partial}\partial + \bar{\partial}^2$$

and decomposing this into types yields $(1.1.93)$, $(1.1.94)$. $\qquad \qquad \square$

1.1.3 Riemannian and Lorentzian Metrics

In local coordinates $x = (x^1, \ldots, x^d)$, a metric is represented by a nondegenerate, symmetric matrix

$$(g_{ij}(x))_{i,j=1,\ldots,d} \qquad (1.1.95)$$

smoothly depending on x. Being symmetric, this matrix has d real eigenvalues, and being nondegenerate, none of them is 0. When they are all positive, the metric is called Riemannian. When only one is positive, and therefore $d - 1$ ones are negative, it is called Lorentzian.[4] The prototype of a Riemannian manifold is Euclidean space, $\mathbb{R}^d$ equipped with its Euclidean metric; the model for a Lorentz manifold is Minkowski space, namely $\mathbb{R}^d$ equipped with the inner product

$$\langle x, y \rangle = x^0 y^0 - x^1 y^1 - \cdots - x^{d-1} y^{d-1}$$

for $x = (x^0, x^1, \ldots, x^{d-1})$, $y = (y^0, y^1, \ldots, y^{d-1})$. (It is customary to use the indices $0, \ldots, d - 1$ in place of $1, \ldots, d$ in the Lorentzian case, in order to better distinguish the time direction corresponding to 0 from the spatial ones.) This space is often denoted by $\mathbb{R}^{1,d-1}$.

The product of two tangent vectors $v, w \in T_p M$ with coordinate representations $(v^1, \ldots, v^d)$ and $(w^1, \ldots, w^d)$ (i.e. $v = v^i \frac{\partial}{\partial x^i}$, $w = w^j \frac{\partial}{\partial x^j}$) is then, as in (1.1.21),

$$\langle v, w \rangle := g_{ij}(x(p)) v^i w^j. \tag{1.1.96}$$

In particular, $\langle \frac{\partial}{\partial x^i}, \frac{\partial}{\partial x^j} \rangle = g_{ij}$. In a Lorentzian manifold, a vector v with $\langle v, v \rangle > 0$ is called time-like, one with $\langle v, v \rangle < 0$ space-like, and a nontrivial one with $\|v\| = 0$ light-like.

A (smooth) curve $\gamma : [a, b] \to M$ ($[a, b]$ a closed interval in $\mathbb{R}$) is called time-like when $\langle \dot{\gamma}(t), \dot{\gamma}(t) \rangle > 0$ for all $t \in [a, b]$. Light- or space-like curves are defined analogously.

Similarly, the length or norm of v is given by

$$\|v\| := \langle v, v \rangle^{\frac{1}{2}} \tag{1.1.97}$$

if $\langle v, w \rangle \geq 0$, and

$$\|v\| := -(-\langle v, v \rangle)^{\frac{1}{2}} \tag{1.1.98}$$

if $\langle v, w \rangle < 0$. On a Riemannian manifold, of course all vectors $v \neq 0$ have positive length.

Starting from the product (1.1.96), a metric then also induces products on other tensors. For example, for cotangent vectors $\omega = \omega_i dx^i$, $\lambda = \lambda_i dx^i \in T_p^* M$, we have

$$\langle \omega, \lambda \rangle = g^{ij}(x(p)) \omega_i \lambda_j, \tag{1.1.99}$$

---

[4]The conventions are not generally agreed upon in the literature (see [81] for a systematic survey of the older literature). The one employed here seems to be the one followed by the majority of physicists. Sometimes, however, for a Lorentzian metric, one requires $d - 1$ positive and 1 negative eigenvalues. Of course, this simply changes the convention adopted here by a minus sign, without affecting the geometric or physical content. The latter convention looks natural when one wants to add a temporal dimension to already present spatial ones. The convention adopted here, in contrast, is natural when one starts with kinetics described by ordinary differential equations derived from a positive definite Lagrangian. Thus, the temporal dimension is the primary one and counted positively, whereas the additional spatial ones then lead to field theories.

that is, the induced product on the cotangent space is given by the inverse of the metric tensor. As a check, the reader should verify that this expression is invariant under coordinate transformations, with the transformation behavior of the metric now presented (or recalled from (1.1.20)).

Let $y = f(x)$. v and w then have representations $(\tilde{v}^1, \ldots, \tilde{v}^d)$ and $(\tilde{w}, \ldots, \tilde{w}^d)$ with $\tilde{v}^j = v^i \frac{\partial f^j}{\partial x^i}, \tilde{w}^j = w^i \frac{\partial f^j}{\partial x^i}$. The metric in the new coordinates, denoted by $h_{k\ell}(y)$, then satisfies

$$h_{k\ell}(f(x))\tilde{v}^k\tilde{w}^\ell = \langle v, w \rangle = g_{ij}(x)v^i w^j. \tag{1.1.100}$$

Therefore, the transformation rule is the one given in (1.1.20),

$$h_{k\ell}(f(x))\frac{\partial f^k}{\partial x^i}\frac{\partial f^\ell}{\partial x^j} = g_{ij}(x). \tag{1.1.101}$$

Given a metric $(g_{ij}(x))_{i,j=1,\ldots,d}$, we put

$$\sqrt{g} := \sqrt{\det(g_{ij})} \tag{1.1.102}$$

and define the Laplace–Beltrami operator (Laplacian for short) acting on $C^\infty(M)$ as

$$\Delta := \Delta_g := \frac{1}{\sqrt{g}}\frac{\partial}{\partial x^i}\left(\sqrt{g}g^{ij}\frac{\partial}{\partial x^j}\right). \tag{1.1.103}$$

We assume that our manifold M is compact (and, as always, without boundary). We then have the integration by parts formula, using $\langle ., . \rangle$ for the product on 1-forms induced by the Riemannian metric g,

$$\int \langle df, dg \rangle \sqrt{g}dx^1 \cdots dx^d = \int g^{ij}\frac{\partial f}{\partial x^i}\frac{\partial g}{\partial x^j}\sqrt{g}dx^1 \cdots dx^d$$
$$= -\int f\Delta g \sqrt{g}dx^1 \cdots dx^d \tag{1.1.104}$$

where $\sqrt{g}dx^1 \ldots dx^d$ is the volume form $dvol_g$ for the Riemannian metric as defined in (1.1.26). (Note that we are always assuming that our manifold M is oriented. This avoids sign ambiguities in the volume form and permits global integration as in (1.1.104).)

In the Euclidean case, the Laplacian is simply the sum of the pure second derivatives,

$$\Delta = \sum_{i=1}^{d}\frac{\partial^2}{(\partial x^i)^2} \tag{1.1.105}$$

(cf. also (1.1.68) above). For the Minkowski metric, we have

$$\Delta = \frac{\partial^2}{\partial(x^0)^2} - \sum_{i=1}^{d-1}\frac{\partial^2}{\partial(x^i)^2}, \tag{1.1.106}$$

and this operator is often denoted by $\Box$ in the literature.

Generalizing (1.1.98), the metric g induces a product $\langle \omega, v \rangle$ on p-forms, see (1.1.25), and we can then define the formal adjoint d^* of the exterior derivative d via

$$\int \langle d\mu, v \rangle dvol_g = \int \langle \mu, d^*v \rangle \, dvol_g \qquad (1.1.107)$$

for a $(p-1)$-form μ and a p-form v. (Since $d : \Omega^p(M) \to \Omega^{p+1}(M)$, i.e., d maps p-forms to $(p+1)$-forms, $d^* : \Omega^{p+1}(M) \to \Omega^p(M)$ maps $(p+1)$-forms to p-forms.) On functions, we then have

$$\Delta f = -d^*df. \qquad (1.1.108)$$

More generally, one defines the Hodge Laplacian on p-forms by

$$dd^* + d^*d. \qquad (1.1.109)$$

Since $d^*f = 0$ for functions, i.e, 0-forms f (for the simple reason that there do not exist forms of degree -1), this is a generalization of (1.1.108)—up to the sign, and these differing sign conventions unfortunately cause a lot of confusion. We then have the general integration by parts formulae for p-forms

$$\int \langle d^*d\mu, v \rangle \, dvol_g = \int \langle d\mu, dv \rangle \, dvol_g = \int \langle \mu, d^*dv \rangle \, dvol_g \qquad (1.1.110)$$

and

$$\int \langle (dd^* + d^*d)\mu, v \rangle \, dvol_g = \int (\langle d\mu, dv \rangle + \langle d^*\mu, d^*v \rangle) \, dvol_g$$

$$= \int \langle \mu, (dd^* + d^*d)v \rangle \, dvol_g. \qquad (1.1.111)$$

Let us briefly explain the relation with the cohomology of the (compact, oriented) manifold M. A p-form ω is called closed if

$$d\omega = 0, \qquad (1.1.112)$$

and it is called exact if there exists some $(p-1)$-form η with

$$\omega = d\eta. \qquad (1.1.113)$$

Because of $d \circ d = 0$, see (1.1.37), any exact form is closed. Two closed p-forms ω^1, ω^2 are considered as cohomologically equivalent if their difference is exact, i.e., if there exists some $(p-1)$-form η with

$$\omega^1 - \omega^2 = d\eta. \qquad (1.1.114)$$

The equivalence classes of p-forms constitute a group, the pth (de Rham) cohomology group $H^p(M)$ of M. When M carries a Riemannian metric g, one can identify a natural representative for each cohomology class as the unique form μ that minimizes

$$\int_M \langle \omega, \omega \rangle dvol_g \qquad (1.1.115)$$

in its equivalence class. This minimizing form μ is then harmonic in the sense that

$$(dd^* + d^*d)\mu = 0, \qquad (1.1.116)$$

or equivalently (as follows from (1.1.111) and the nonnegativity of the two terms in the middle integral)

$$d\mu = 0 \quad \text{and} \quad d^*\mu = 0. \qquad (1.1.117)$$

Thus, a harmonic form is closed ($d\mu = 0$) and coclosed ($d^*\mu = 0$).

Since M is compact, the dimension $b_p(M)$ (called the pth Betti number of M) of $H^p(M)$ is finite. This follows for instance from the fact that the elements of $H^p(M)$ are identified with the solutions of the elliptic differential equation (1.1.116). It is a general result in the theory of elliptic partial differential equations that their solution spaces satisfy a compactness principle.

1.1.4 Geodesics

The length of a smooth (or at least rectifiable) curve $\gamma : [a, b] \to M$ is

$$L(\gamma) := \int_a^b \left\| \frac{d\gamma}{dt}(t) \right\| dt = \int_a^b \sqrt{g_{ij}(x(\gamma(t)))\dot{x}^i(t)\dot{x}^j(t)} dt \qquad (1.1.118)$$

where we abbreviate $\dot{x}^i(t) := \frac{d}{dt}(x^i(\gamma(t)))$. Thus, time-, light-, or space-like curves have positive, vanishing, or negative length, respectively

The action of a time-like curve γ is

$$S(\gamma) := \frac{1}{2} \int_a^b \left\| \frac{d\gamma}{dt}(t) \right\|^2 dt = \frac{1}{2} \int_a^b g_{ij}(x(\gamma(t)))\dot{x}^i(t)\dot{x}^j(t) dt. \qquad (1.1.119)$$

Here, γ is considered as the orbit of a mass point, which explains the name "action". In the mathematical literature, the action is often called energy, an unfortunate choice of terminology.

A massive particle in a Lorentzian manifold travels along a world line $x(\tau)$ with arclength

$$s = \int_{\tau_0}^{\tau_1} \left(g_{\alpha\beta}(x(\tau))\dot{x}^\alpha(\tau)\dot{x}^\beta(\tau) \right)^{\frac{1}{2}} d\tau,$$

where we assume $g_{\alpha\beta}\dot{x}^\alpha\dot{x}^\beta > 0$ along the world line. Thus, the movement is time-like. When in place of $g_{\alpha\beta}\dot{x}^\alpha\dot{x}^\beta > 0$, we have

$$g_{\alpha\beta}\dot{x}^\alpha\dot{x}^\beta = 0,$$

then the particle is massless, that is, a photon. $g_{\alpha\beta}\dot{x}^\alpha\dot{x}^\beta < 0$ would correspond to a movement with speed higher than that of light and is excluded.

By Hölder's inequality, for a time-like curve γ,

$$\int_a^b \left\| \frac{d\gamma}{dt} \right\| dt \le (b-a)^{\frac{1}{2}} \left(\int_a^b \left\| \frac{d\gamma}{dt} \right\|^2 dt \right)^{\frac{1}{2}} \tag{1.1.120}$$

with equality precisely if $\| \frac{d\gamma}{dt} \| \equiv$ const. This means that

$$L(\gamma)^2 \le 2(b-a)S(\gamma), \tag{1.1.121}$$

again with equality only if γ has constant norm.

The distance between $p, q \in M$ is

$$d(p,q) := \inf\{L(\gamma) : \gamma : [a,b] \to M \text{ with } \gamma(a) = p, \gamma(b) = q\}. \tag{1.1.122}$$

By the change of variables formula, if $\gamma : [a,b] \to M$ is a curve, and $\sigma : [a', b'] \to [a, b]$ is a change of parameter, then

$$L(\gamma \circ \sigma) = L(\gamma). \tag{1.1.123}$$

This is no longer so for the action, as follows with a little reflection on the equality discussion in (1.1.121). It is instructive to look at the stationary points of the action:

Lemma 1.3 *The Euler–Lagrange equations (see Sect. 2.3.1 below) for the action S are*

$$\ddot{x}^i(t) + \Gamma^i_{jk}(x(t))\dot{x}^j(t)\dot{x}^k(t) = 0, \quad i = 1,\dots,d, \tag{1.1.124}$$

where Γ^i_{jk} are the Christoffel symbols (1.1.60).

Proof As will be derived in Sect. 2.3.1 below, the Euler–Lagrange equations of a functional

$$I(x) = \int_a^b f(t, x(t), \dot{x}(t))dt$$

are given by

$$\frac{d}{dt}\frac{\partial f}{\partial \dot{x}^i} - \frac{\partial f}{\partial x^i} = 0, \quad i = 1,\dots,d.$$

Thus, for our action S,

$$\frac{d}{dt}(g_{ik}(x(t))\dot{x}^k(t) + g_{ji}(x(t))\dot{x}^j(t)) - g_{jk,i}(x(t))\dot{x}^j(t)\dot{x}^k(t) = 0$$

for $i = 1,\dots,d,$

hence

$$g_{ik}\ddot{x}^k + g_{ji}\ddot{x}^j + g_{ik,\ell}\dot{x}^\ell \dot{x}^k + g_{ji,\ell}\dot{x}^\ell \dot{x}^j - g_{jk,i}\dot{x}^j \dot{x}^k = 0.$$

Renaming indices and using $g_{ik} = g_{ki}$, we get

$$2g_{\ell m}\ddot{x}^m + (g_{\ell k,j} + g_{j\ell,k} - g_{jk,\ell})\dot{x}^j \dot{x}^k = 0$$

and from this

$$g^{i\ell} g_{\ell m}\ddot{x}^m + \frac{1}{2}g^{i\ell}(g_{\ell k,j} + g_{j\ell,k} - g_{jk,\ell})\dot{x}^j \dot{x}^k = 0.$$

Because of $g^{i\ell} g_{\ell m} = \delta^i_m$ and thus $g^{i\ell} g_{\ell m}\ddot{x}^m = \ddot{x}^i$, (1.1.124) follows. $\square$

Definition 1.3 A geodesic is a curve $\gamma = [a, b] \to M$ that is a critical point of the action S, that is, satisfies (1.1.124).

Briefly interrupting our discussion, we point out that (1.1.124) is the same as (1.1.55). In other words, taking up the discussion at the end of Sect. 1.1.1, for the Levi-Cività connection, the two definitions of a geodesic, being autoparallel as in Sect. 1.1.1, or being a critical point of the action functional S as defined here, are equivalent. In particular, we can also write the geodesic equation invariantly, as in (1.1.54), with a slight change of notation:

$$\nabla_{\frac{d}{dt}}\dot{x} = 0. \tag{1.1.125}$$

We now return to the discussion of geodesics as critical points of S. We say that a curve γ is parametrized proportionally to arc length if $\langle \dot{x}, \dot{x} \rangle \equiv$ const.

Lemma 1.4 *Each geodesic is parametrized proportionally to arc length.*

Proof For a solution of (1.1.124),

$$\frac{d}{dt}\langle \dot{x}, \dot{x} \rangle = \frac{d}{dt}(g_{ij}(x(t))\dot{x}^i(t)\dot{x}^j(t))$$

$$= g_{ij}\ddot{x}^i\dot{x}^j + g_{ij}\dot{x}^i\ddot{x}^j + g_{ij,k}\dot{x}^i\dot{x}^j\dot{x}^k$$

$$= -(g_{jk,\ell} + g_{\ell j,k} - g_{\ell k,j})\dot{x}^\ell\dot{x}^k\dot{x}^j + g_{\ell j,k}\dot{x}^k\dot{x}^\ell\dot{x}^j$$

$$= 0, \quad \text{since } g_{jk,\ell}\dot{x}^\ell\dot{x}^k\dot{x}^j = g_{\ell k,j}\dot{x}^\ell\dot{x}^k\dot{x}^j.$$

Consequently, $\langle \dot{x}, \dot{x} \rangle \equiv$ const., and hence the curve is parametrized proportionally to arc length. $\square$

As already discussed in Sect. 1.1.1, the next result follows from the Picard–Lindelöf theorem about the local existence and uniqueness of solutions of systems of ODEs.

Lemma 1.5 *For each $p \in M$, $v \in T_p M$, there exist $\varepsilon > 0$ and precisely one geodesic*

$$c : [0, \varepsilon] \to M$$

with $c(0) = p$ and $\dot{c}(0) = v$. This geodesic c depends smoothly on p and v.

We now assume that the metric g on M is Riemannian, even though results corresponding to those stated below also hold in the case of other signatures, in particular for Lorentzian metrics.

If $x(t)$ is a solution of (1.1.124), so is $x(\lambda t)$ for any constant $\lambda \in \mathbb{R}$. Denoting the geodesic of Lemma 1.5 by c_v,

$$c_v(t) = c_{\lambda v}\left(\frac{t}{\lambda}\right) \quad \text{for } \lambda > 0, \ t \in [0, \varepsilon].$$

In particular, $c_{\lambda v}$ is defined on $[0, \frac{\varepsilon}{\lambda}]$.

Since c_v depends smoothly on v, and $\{v \in T_p M : \|v\| = 1\}$ is compact, there exists $\varepsilon_0 > 0$ with the property that for $\|v\| = 1$, c_v is defined at least on $[0, \varepsilon_0]$. Therefore, for any $w \in T_p M$ with $\|w\| \leq \varepsilon_0$, c_w is defined at least on $[0, 1]$. Thus, as in (1.1.56):

Definition 1.4 Let $p \in M$, $V_p := \{v \in T_p M : c_v \text{ is defined on } [0, 1]\}$.

$$\exp_p : V_p \to M,$$

$$v \mapsto c_v(1) \tag{1.1.126}$$

is called the exponential map of M at p.

One observes that the derivative of the exponential map $\exp_p$ at $0 \in T_p M$ is the identity. Therefore, with the help of the inverse function theorem, one checks that the exponential map $\exp_p$ maps a neighborhood of $0 \in T_p M$ diffeomorphically onto a neighborhood of $p \in M$. Since $T_p M$ is a vector space isomorphic to $\mathbb{R}^d$ (on which we choose a Euclidean orthonormal basis), we can consider the local inverse $\exp_p^{-1}$ as defining local coordinates in a neighborhood of p. These local coordinates are called normal coordinates with center p. In these coordinates, a basis of $T_p M$ that is orthonormal with respect to the Riemannian metric g is identified with a Euclidean orthonormal basis of $\mathbb{R}^d$. This is the first part of the next lemma:

Lemma 1.6 *In normal coordinates, the metric satisfies*

$$g_{ij}(0) = \delta_{ij}, \tag{1.1.127}$$

$$\Gamma^i_{jk}(0) = 0 \quad (\text{and also } g_{ij,k}(0) = 0) \quad \text{for all } i, j, k. \tag{1.1.128}$$

Proof (1.1.127) follows from the fact that the above identification $\Phi : T_p M \cong \mathbb{R}^d$ maps an orthonormal basis of $T_p M$ w.r.t. the metric g (that is, a basis $e_1, \ldots e_d$ with

$\langle e_i, e_j \rangle = \delta_{ij}$ onto an orthonormal basis of $\mathbb{R}^d$. For (1.1.128), in normal coordinates, the straight lines through the origin of $\mathbb{R}^d$ are geodesic, as the line tv, $t \in \mathbb{R}$, $v \in \mathbb{R}^d$, is mapped onto $c_{tv}(1) = c_v(t)$, where $c_v(t)$ is the geodesic, parametrized by arc length, with $\dot{c}_v(0) = v$.

Inserting now $x(t) = tv$ into the geodesic equation (1.1.124), we obtain, using $\ddot{x}(t) = 0$,

$$\Gamma^i_{jk}(tv)v^j v^k = 0, \quad \text{for } i = 1, \ldots, d.$$

In particular at 0, i.e., for $t = 0$,

$$\Gamma^i_{jk}(0)v^j v^k = 0 \quad \text{for all } v \in \mathbb{R}^d, \ i = 1, \ldots, d.$$

Using the symmetry $\Gamma^i_{jk} = \Gamma^i_{kj}$, this implies

$$\Gamma^i_{\ell m}(0) = 0$$

for all i and also for all ℓ, m. By definition of Γ^i_{jk}, at $0 \in \mathbb{R}^d$, we obtain

$$g^{i\ell}(g_{j\ell,k} + g_{k\ell,j} - g_{jk,\ell}) = 0$$

for all free indices, hence also

$$g_{jm,k} + g_{km,j} - g_{jk,m} = 0.$$

Permuting the indices yields

$$g_{kj,m} + g_{mj,k} - g_{km,j} = 0,$$

which we add to obtain, for all indices,

$$g_{jm,k}(0) = 0. \qquad \square$$

This is a very useful result. When one has to check tensor equations, one can do this in arbitrary coordinates because by the definition of a tensor, results are coordinate independent. Now, it is often much easier to check such identities in normal coordinates at the point under consideration, making use of the vanishing of all first derivatives of the metric and all Christoffel symbols. We shall often employ this strategy in the sequel.

In fact, we can even achieve a little more: Let $c(s) : (-a, a) \to M$ be a geodesic parametrized by arclength, that is, $\langle \dot{c}(s), \dot{c}(s) \rangle = 1$ for $-a < s < a$ (see Lemma 1.4). Let $v^1(0), \ldots, v^d(0)$ be an orthonormal basis of $T_{c(0)}M$ with $v^1 = \dot{c}(0)$, and let $v^i(t) \in T_{c(t)}M$ be the parallel transport of $v^i(0)$ along the geodesic $c(s)$. We define coordinates by mapping $(x^1, \ldots, x^d)$ in some neighborhood of $0 \in \mathbb{R}^d$ to

$$(c(x^1), \exp_{c(x^1)}(x^2 v^2(x^1) + \cdots + x^d v^d(x^1))). \tag{1.1.129}$$

Lemma 1.7 *The coordinates just described satisfy*

$$g_{ij}(x^1, 0, \ldots, 0) = \delta_{ij}, \qquad (1.1.130)$$

$$\Gamma^i_{jk}(x^1, 0, \ldots, 0) = 0, \qquad (1.1.131)$$

$$(\text{and also } g_{ij,k}(x^1, 0, \ldots, 0) = 0) \qquad (1.1.132)$$

for all $-a < x^1 < a, i, j, k.$

Proof By Lemma 1.4, $g_{11}(x^1, 0, \ldots, 0)$ is constant, in fact $\equiv 1$ by our arclength assumption, as a function of x^1. Therefore, also $g_{11,1}(x^1, 0, \ldots, 0) = 0$. Moreover, since the Levi-Città connection ∇ respects the metric (see (1.1.62)), $g_{jk}(x^1, 0, \ldots, 0)$
$= \langle v^i(x^1), v^j(x^1) \rangle = \delta_{jk}$ for the other values of j, k. Therefore, also

$$g_{jk,1}(x^1, 0, \ldots, 0) = 0 \quad \text{for all } j, k. \qquad (1.1.133)$$

We continue to evaluate all expressions at $(x^1, 0, \ldots, 0)$. All rays tv for v in the span of $v^2, \ldots, v^d$ are mapped to geodesics, because the exponential map is applied to them. So, we obtain, as in the proof of Lemma 1.6, that

$$\Gamma^i_{\ell m}(x^1, 0, \ldots, 0) = 0$$

for $i = 2, \ldots, d$ and all ℓ, m. By definition of Γ^i_{jk}, we obtain $g^{i\ell}(g_{j\ell,k} + g_{k\ell,j} - g_{jk,\ell}) = 0$ at $(x^1, 0, \ldots, 0) \in \mathbb{R}^d$ for all free indices, hence also $g_{jm,k} + g_{km,j} - g_{jk,m} = 0$ for $m = 2, \ldots, d$. Permuting the indices to get $g_{kj,m} + g_{mj,k} - g_{km,j} = 0$, adding these relations and combining them with (1.1.133) finally yields $g_{jm,k}(x^1, 0, \ldots, 0) = 0$ for all indices. $\qquad \square$

1.1.5 Curvature

We now want to discuss the curvature tensor R of the Levi-Città connection ∇. We recall (1.1.51):

$$R(X, Y)Z = \nabla_X \nabla_Y Z - \nabla_Y \nabla_X Z - \nabla_{[X,Y]}Z. \qquad (1.1.134)$$

In local coordinates (cf. (1.1.52)),

$$R\left(\frac{\partial}{\partial x^i}, \frac{\partial}{\partial x^j}\right)\frac{\partial}{\partial x^\ell} = R^k_{\ell ij}\frac{\partial}{\partial x^k}. \qquad (1.1.135)$$

We put

$$R_{k\ell ij} := g_{km} R^m_{\ell ij}, \qquad (1.1.136)$$

i.e.

$$R_{k\ell ij} = \left\langle R\left(\frac{\partial}{\partial x^i}, \frac{\partial}{\partial x^j}\right) \frac{\partial}{\partial x^\ell}, \frac{\partial}{\partial x^k}\right\rangle. \tag{1.1.137}$$

There exist different sign conventions for the curvature tensor in the literature. We have adopted here a convention that hopefully minimizes conflict between them. As a consequence, the indices k and l appear in different orders at the two sides of (1.1.137).

The curvature tensor satisfies the following symmetries:

$$R(X, Y)Z = -R(Y, X)Z, \quad \text{i.e.} \quad R_{k\ell ij} = -R_{k\ell ji} \tag{1.1.138}$$

for vector fields X, Y, Z, W.

$$R(X, Y)Z + R(Y, Z)X + R(Z, X)Y = 0, \tag{1.1.139}$$

or with indices

$$R_{k\ell ij} + R_{kij\ell} + R_{kj\ell i} = 0 \tag{1.1.140}$$

(the first Bianchi identity).

$$\langle R(X, Y)Z, W\rangle = -\langle R(X, Y)W, Z\rangle, \tag{1.1.141}$$

with indices

$$R_{k\ell ij} = -R_{\ell kij}. \tag{1.1.142}$$

$$\langle R(X, Y)Z, W\rangle = \langle R(Z, W)X, Y\rangle, \tag{1.1.143}$$

with indices

$$R_{k\ell ij} = R_{ijk\ell}. \tag{1.1.144}$$

$$\frac{\partial}{\partial x^h} R_{k\ell ij} + \frac{\partial}{\partial x^k} R_{\ell hij} + \frac{\partial}{\partial x^\ell} R_{hkij} = 0 \tag{1.1.145}$$

(the second Bianchi identity). In order to practice tensor calculus, we give a proof of (1.1.145) in local coordinates. We recall (1.1.53):

$$R_{lij}^k = \frac{\partial}{\partial x^i} \Gamma_{jl}^k - \frac{\partial}{\partial x^j} \Gamma_{il}^k + \Gamma_{im}^k \Gamma_{jl}^m - \Gamma_{jm}^k \Gamma_{il}^m. \tag{1.1.146}$$

Since all expressions are tensors, we may choose normal coordinates around the point x_0 under consideration, i.e., for all indices

$$g_{ij}(x_0) = \delta_{ij}, \; g_{ij,k}(x_0) = 0 = \Gamma_{ij}^k(x_0) \tag{1.1.147}$$

(1.1.146) then becomes

$$R_{k\ell ij} = \frac{1}{2}(g_{jk,\ell i} + g_{\ell k,ij} - g_{j\ell,ki} - g_{ik,\ell j} - g_{\ell k,ij} + g_{i\ell,kj})$$

$$= \frac{1}{2}(g_{jk,\ell i} + g_{i\ell,kj} - g_{j\ell,ki} - g_{ik,\ell j}), \quad (1.1.148)$$

and also, differentiating (1.1.146) and using once more the vanishing of all terms containing first derivatives of g_{ij} at x_0,

$$R_{k\ell ij,h} = \frac{1}{2}(g_{jk,\ell ih} + g_{i\ell,kjh} - g_{j\ell,kih} - g_{ik,\ell jh}). \quad (1.1.149)$$

This yields the second Bianchi identity:

$$R_{k\ell ij,h} + R_{\ell hij,k} + R_{hkij,\ell} = \frac{1}{2}(g_{jk,\ell ih} + g_{i\ell,kjh} - g_{j\ell,kih} - g_{ik,\ell jh}$$

$$+ g_{j\ell,hik} + g_{ih,\ell jk} - g_{jh,\ell ik} - g_{i\ell,hjk}$$

$$+ g_{jh,ki\ell} + g_{ik,hj\ell} - g_{jk,hi\ell} - g_{ih,kj\ell})$$

$$= 0.$$

The sectional curvature of the plane spanned by the (linearly independent) tangent vectors $X = \xi^i \frac{\partial}{\partial x^i}, Y = \eta^i \frac{\partial}{\partial x^i} \in T_x M$ is defined as

$$K(X \wedge Y) := \langle R(X, Y)Y, X \rangle \frac{1}{|X \wedge Y|^2}, \quad (1.1.150)$$

$(|X \wedge Y|^2 = \langle X, X \rangle \langle Y, Y \rangle - \langle X, Y \rangle^2)$, with indices

$$K(X \wedge Y) = \frac{R_{ijk\ell}\xi^i \eta^j \xi^k \eta^\ell}{g_{ik}g_{j\ell}(\xi^i \xi^k \eta^j \eta^\ell - \xi^i \xi^j \eta^k \eta^\ell)} = \frac{R_{ijk\ell}\xi^i \eta^j \xi^k \eta^\ell}{(g_{ik}g_{j\ell} - g_{ij}g_{k\ell})\xi^i \eta^j \xi^k \eta^\ell}. \quad (1.1.151)$$

The Ricci curvature in the direction $X = \xi^i \frac{\partial}{\partial x^i} \in T_x M$ is defined as the average of the sectional curvatures of all planes in $T_x M$ containing X,

$$\mathrm{Ric}(X, X) = g^{j\ell}\left\langle R\left(X, \frac{\partial}{\partial x^j}\right)\frac{\partial}{\partial x^\ell}, X \right\rangle, \quad (1.1.152)$$

and the Ricci tensor is thus the contraction of the curvature tensor,

$$R_{ik} = g^{j\ell}R_{ijk\ell} = R^j_{ijk}. \quad (1.1.153)$$

(1.1.144) implies the symmetry

$$R_{ik} = R_{ki}. \quad (1.1.154)$$

The scalar curvature is the contraction of the Ricci curvature,

$$R = g^{ik}R_{ik} = R^i_i. \quad (1.1.155)$$

For $d = \dim M = 2$, the curvature tensor is determined by the scalar curvature:

$$R_{ijk\ell} = R(g_{ik}g_{j\ell} - g_{ij}g_{k\ell}). \quad (1.1.156)$$

For $d = 3$, the curvature tensor is determined by the Ricci tensor. For $d > 3$, the part of the curvature tensor not yet determined by the Ricci tensor is given by the Weyl tensor

$$W_{ijk\ell} = R_{ijk\ell} + \frac{2}{d-2}(g_{i\ell}R_{kj} - g_{ik}R_{\ell j} + g_{jk}R_{\ell i} - g_{j\ell}R_{ki})$$

$$+ \frac{2}{(d-1)(d-2)}R(g_{ik}g_{\ell j} - g_{i\ell}g_{kj}). \tag{1.1.157}$$

1.1.6 Principles of General Relativity

General relativity describes the physical force of gravity and its relation with the structure of space–time. The fundamental physical insight behind the theory of general relativity is that the effects of acceleration cannot be distinguished from those of gravity. The presence of matter changes the geometry of space, and acceleration is experienced in relation to that geometry. In particular, the geometry of space and time is dynamically determined by the physical laws, and in contrast to other physical theories, is thus not assumed as independently given. These physical laws in turn are deduced from symmetry principles, more precisely from the principle of general covariance, that is, that the physics should be independent of its coordinate description. For this, Riemannian geometry has developed the appropriate formal tools.

Let M be a Lorentz manifold with local coordinates (x^0, x^1, x^2, x^3) and metric

$$(g_{\alpha\beta})_{\alpha,\beta=0,...,3}.$$

We recall the Christoffel symbols

$$\Gamma^\alpha_{\beta\gamma} = \frac{1}{2}g^{\alpha\delta}(g_{\beta\delta,\gamma} + g_{\gamma\delta,\beta} - g_{\beta\gamma,\delta}),$$

and those objects from which the essential invariants of a metric come, that is, the curvature tensor $R^\alpha_{\beta\gamma\delta} = \Gamma^\alpha_{\beta\delta,\gamma} - \Gamma^\alpha_{\beta\gamma,\delta} + \Gamma^\alpha_{\eta\gamma}\Gamma^\eta_{\beta\delta} - \Gamma^\alpha_{\eta\delta}\Gamma^\eta_{\beta\gamma}$, and its contractions, the Ricci tensor (1.1.154), $R_{\alpha\beta} = R^\gamma_{\alpha\gamma\beta}$, and the scalar curvature (1.1.156), $R = g^{\alpha\beta}R_{\alpha\beta}$.

The Einstein field equations couple the metric $g_{\alpha\beta}$ of the underlying differentiable manifold with the matter and fields on that manifold. These equations involve the Ricci curvature and are

$$R^{\alpha\beta} - \frac{1}{2}g^{\alpha\beta}R = \kappa T^{\alpha\beta}. \tag{1.1.158}$$

Here, $\kappa = \frac{8\pi g}{c^4}$ where g is the gravitational constant. $(T^{\alpha\beta})_{\alpha,\beta}$ is the energy–momentum tensor. It describes the matter and fields present. When T is given, the Einstein equations then determine the metric of space–time.[5] The presence of a nonvanishing energy–momentum tensor in the field equations makes space–time curved. The curvature in turn leads to gravity. (1.1.158) is equivalent to

$$R_{\alpha\beta} - \frac{1}{2} g_{\alpha\beta} R = \kappa T_{\alpha\beta}. \qquad (1.1.159)$$

Taking the trace in (1.1.159) leads to $R - 2R = \kappa T$, that is,

$$R = -\kappa T. \qquad (1.1.160)$$

($T = g^{\alpha\beta} T_{\alpha\beta}$; note that the dimension of M is 4.) Using (1.1.160), (1.1.159) becomes equivalent to

$$R_{\alpha\beta} = \kappa \left(T_{\alpha\beta} - \frac{1}{2} g_{\alpha\beta} T \right). \qquad (1.1.161)$$

In the special case where $(T_{\alpha\beta}) = 0$, that is, when neither matter nor fields are present, (1.1.161) becomes

$$R_{\alpha\beta} = 0, \qquad (1.1.162)$$

i.e., the Ricci curvature of M vanishes.

Hilbert discovered that the Einstein field equations can be derived from a variational principle. In fact, they are the Euler–Lagrange equations for the action functional

$$L_0(g) = \int_M R\sqrt{-g}\, dx = \int_M R\, d\text{Vol}_M(x), \qquad (1.1.163)$$

called the Einstein–Hilbert functional. To see this, we consider a family

$$g_{\alpha\beta}^t = g_{\alpha\beta} + t h_{\alpha\beta}$$

of metrics with $(h_{\alpha\beta})$ having compact support if M is not compact itself. Quantities obtained from the metric $g_{\alpha\beta}^t$ will always carry a superscript t; for example,

$$R_{\alpha\beta}^t$$

is the Ricci tensor of $g_{\alpha\beta}^t$. We also put

$$\delta g_{\alpha\beta} = \frac{d}{dt}(g_{\alpha\beta}^t)_{|t=0} = h_{\alpha\beta},$$

$$\delta R_{\alpha\beta} = \frac{d}{dt}(R_{\alpha\beta}^t)_{|t=0}, \quad \text{etc.}$$

---

[5]Classically, the topology of M is assumed fixed. However, it turns out that the equations may lead to space–time singularities, like black holes, which will then affect the underlying topology. Such singularities can occur and are sometimes even inevitable, even if suitable and physically natural restrictions are imposed on the energy–momentum tensor, like nonnegativity. We do not pursue that issue here, however, but refer to [56]. There, also the cosmological implications of such singularities are discussed.

Finally, we shall use the abbreviation

$$\gamma := \sqrt{-g}.$$

We then have

$$\frac{d}{dt} L_0(g^t)_{|t=0} = \int_M \delta(R\gamma) \, dx. \tag{1.1.164}$$

Now

$$\delta(R\gamma) = \delta(g^{\alpha\beta} R_{\alpha\beta}\gamma) = g^{\alpha\beta}\gamma\delta R_{\alpha\beta} + R_{\alpha\beta}\delta(g^{\alpha\beta}\gamma).$$

We now claim that

$$g^{\alpha\beta}\delta R_{\alpha\beta} = \operatorname{div} V \quad \left(= \frac{1}{\gamma}\frac{\partial}{\partial x^\alpha}(\gamma V^\alpha) \right) \tag{1.1.165}$$

for the vector field V with components

$$V^\gamma = g^{\alpha\beta}\delta\Gamma^\gamma_{\alpha\beta} - g^{\gamma\alpha}\delta\Gamma^\beta_{\beta\alpha}. \tag{1.1.166}$$

Proof of (1.1.165): Let $p \in M$. We introduce normal coordinates near p; thus, at p, the metric tensor is diagonal and

$$g_{\alpha\beta,\gamma}(p) = 0 \quad \text{and} \quad \Gamma^\alpha_{\beta\gamma}(p) = 0 \quad \text{for all } \alpha, \beta, \gamma.$$

In particular, at p

$$\frac{\partial}{\partial x^\alpha}\gamma = 0 \quad \text{for all } \alpha.$$

In these coordinates, (1.1.165) then follows from the definition of the Ricci tensor. While the Christoffel symbols $\Gamma^\alpha_{\beta\gamma}$, as the components of a connection, do not transform tensorially, the $\delta\Gamma^\alpha_{\beta\gamma}$ do transform tensorially as derivatives, that is, as infinitesimal differences of connections. The right-hand side of (1.1.165) is thus a tensor, and so is the left-hand side. The equality of two tensors can be checked in arbitrary coordinates. Since we have just verified (1.1.165) in normal coordinates, (1.1.165) then also holds in arbitrary coordinates, and we have completed its proof.

We now get

$$\int_M \delta(R\gamma) = \int g^{\alpha\beta}\gamma\delta R_{\alpha\beta} \, dx + \int R_{\alpha\beta}\delta(g^{\alpha\beta}\gamma) \, dx$$

$$= \int \operatorname{div} V \gamma \, dx + \int R_{\alpha\beta}\delta(g^{\alpha\beta}\gamma) \, dx$$

$$= \int R_{\alpha\beta}\delta(g^{\alpha\beta}\gamma) \, dx \tag{1.1.167}$$

by Gauss's theorem, since V has compact support.

Now

$$\delta\gamma = -\frac{1}{2}\gamma^{-1}\delta\det(g_{\alpha\beta}) = \frac{1}{2}\gamma g^{\alpha\beta}\delta g_{\alpha\beta}$$

and moreover

$$\delta g^{\alpha\beta} = -g^{\alpha\gamma}g^{\beta\gamma}\delta g_{\gamma\delta} \quad (\text{from } g^{\alpha\beta}g_{\beta\gamma} = \delta^{\alpha}_{\gamma})$$

and therefore

$$\delta(g^{\alpha\beta}\gamma) = \gamma\left(\frac{1}{2}g^{\alpha\beta}g^{\gamma\delta} - g^{\alpha\gamma}g^{\beta\delta}\right)\delta g_{\gamma\delta}. \tag{1.1.168}$$

(1.1.164), (1.1.167), (1.1.168) imply

$$\delta L_0 = \int_M \left(\frac{1}{2}g^{\alpha\beta}R - R^{\alpha\beta}\right)\gamma\delta g_{\alpha\beta}\, dx = 0. \tag{1.1.169}$$

If this holds for all variations $\delta g_{\alpha\beta}$ with compact support, we have

$$R^{\alpha\beta} - \frac{1}{2}g^{\alpha\beta}R = 0, \tag{1.1.170}$$

which implies, as in the derivation of (1.1.162), that

$$R_{\alpha\beta} = 0. \tag{1.1.171}$$

Einstein also tentatively introduced a cosmological constant Λ that has the effect of changing the Einstein–Hilbert functional (1.1.163) to

$$L_\Lambda(g) = \int_M (R - 2\Lambda)\sqrt{-g}\, dx \tag{1.1.172}$$

and the Einstein field equations (1.1.158) to

$$R^{\alpha\beta} - \frac{1}{2}g^{\alpha\beta}R + \Lambda g^{\alpha\beta} = \kappa T^{\alpha\beta}. \tag{1.1.173}$$

While a nontrivial cosmological constant is presently appearing in some cosmological models, we put it to 0 for our present discussion. It is straightforward, however, to include a nontrivial Λ in the subsequent formulas.

In the presence of some matter fields ϕ, we assume a Lagrangian

$$L_1 = \int_M F(g, \phi, \nabla^g \phi)\sqrt{-g}\, dx \tag{1.1.174}$$

depending on the fields and their covariant derivatives w.r.t. the Levi-Cività connection, as well as possibly also directly on the metric g. When we consider a variation $\delta g_{\alpha\beta}$ of the metric that does not change the fields, we put

$$\delta L_1 = \frac{1}{2}\int_M T^{\alpha\beta}\delta g_{\alpha\beta} dx. \tag{1.1.175}$$

In other words, the energy–momentum tensor is defined as the variation of the matter Lagrangian w.r.t. the metric. In order to fully justify (1.1.175), we need to observe that all the variations of all metric dependent terms in L_1 are proportional to $\delta g_{\alpha\beta}$. For the volume form, this has been verified in (1.1.168). The covariant derivative ∇^g occurring in (1.1.175) also depends on the metric (see (1.1.60)). One easily computes, for example in normal coordinates, that the variation $\delta\Gamma^\alpha_{\beta\delta}$ of the Christoffel symbol is proportional to a combination of covariant derivatives of $\delta g_{\alpha\beta}$, and that the covariant derivatives can then be integrated away by parts in the computation of δL_1. When one then considers the full Lagrangian

$$L := L_0 + \kappa L_1, \tag{1.1.176}$$

with κ as a coupling constant, we thus obtain from (1.1.169) and (1.1.175), for variations $\delta g_{\alpha\beta}$ of the metric,

$$\delta L = \int_M \left(\frac{1}{2} g^{\alpha\beta} R - R^{\alpha\beta} + \kappa T^{\alpha\beta} \right) \gamma \delta g_{\alpha\beta} \, dx. \tag{1.1.177}$$

Thus, when $\delta L = 0$ for all variations of the metric, we obtain (1.1.158).

For a more extended discussion of this variational principle, we refer to the presentation in [81] or [56].

Finally, we mention the so-called semiclassical Einstein equations

$$R_{\alpha\beta} - \frac{1}{2} g_{\alpha\beta} R = \kappa \langle \psi | \hat{T}_{\alpha\beta} | \psi \rangle \tag{1.1.178}$$

where the energy momentum tensor $T_{\alpha\beta}$ in (1.1.159) is replaced by the expectation value of the energy–momentum operator with respect to some quantum state ψ. This quantum state in turn depends on the metric g through the Schrödinger equation. Here, we are invoking concepts that find their natural place in the second part of this book. Equation (1.1.178) arises in the context of quantum fields on an external space–time. The coupling of a quantum system to a classical one in (1.1.178) leads to questions of consistency which we do not enter here. We refer to the discussion in [74].

Variational principles will be taken up in more generality below in Sect. 2.3.1, and in Sect. 2.4, the energy–momentum tensor will appear again. Also, we shall see there that a consequence of Noether's theorem (see Sect. 2.3.2) is that under the fundamental assumption of general relativity, namely invariance of L under coordinate transformations—that is, diffeomorphism invariance—the energy–momentum tensor is divergence free.

1.2 Bundles and Connections

1.2.1 Vector and Principal Bundles

Let M be a differentiable manifold. In this section, we present the basic aspects of the theory of vector and principal bundles. We point out that we have already studied one particular vector bundle over M in Sect. 1.1.1, its tangent bundle TM.

A fiber bundle (or simply, a bundle) over M consists of a total space E, a fiber F (both of them also differentiable manifolds), and a projection $\pi : E \to M$ such that each $x \in M$ has a neighborhood U for which $E_{|U} = \pi^{-1}(U)$ is diffeomorphic to $U \times F$ such that the fibers are preserved. This means that there exists a diffeomorphism

$$\varphi : \pi^{-1}(U) \to U \times F$$

with

$$\pi = p_1 \circ \varphi. \tag{1.2.1}$$

($p_1 : U \times F \to U$ is the projection onto the first factor.)

φ is called a local trivialization of the bundle over U. Let $\{U_\alpha\}$ be an open covering of M with local trivializations $\{\varphi_\alpha\}$. If

$$U_\alpha \cap U_\beta \neq \emptyset,$$

we obtain transition maps

$$\varphi_{\beta\alpha} : U_\alpha \cap U_\beta \to \text{Diff}(F) \quad (= \text{group of diffeomorphisms of } F)$$

via

$$\varphi_\beta \circ \varphi_\alpha^{-1}(x, v) = (x, \varphi_{\beta\alpha}(x)v). \tag{1.2.2}$$

Omitting the base point, which is fixed by (1.2.1), from our notation, we shall usually simply write

$$\varphi_{\beta\alpha} = \varphi_\beta \circ \varphi_\alpha^{-1}.$$

We have

$$\text{for } x \in U_\alpha: \quad \varphi_{\alpha\alpha}(x) = \text{id}_F, \tag{1.2.3}$$

$$\text{for } x \in U_\alpha \cap U_\beta: \quad \varphi_{\alpha\beta}(x)\varphi_{\beta\alpha}(x) = \text{id}_F, \tag{1.2.4}$$

$$\text{for } x \in U_\alpha \cap U_\beta \cap U_\gamma: \quad \varphi_{\alpha\gamma}(x)\varphi_{\gamma\beta}(x)\varphi_{\beta\alpha}(x) = \text{id}_F. \tag{1.2.5}$$

E can be reconstructed from its transition maps:

$$E = \coprod_\alpha U_\alpha \times F / \sim \tag{1.2.6}$$

with

$$(x, v) \sim (y, w) :\Leftrightarrow x = y \quad \text{and} \quad w = \varphi_{\beta\alpha}(x)v$$

$$(x \in U_\alpha, y \in U_\beta, v, w \in F).$$

When we have some (differentiable) $f_\alpha : U_\alpha \to \text{Diff}(F)$, we can replace the trivialization φ_α over U_α by

$$\varphi_\alpha' = f_\alpha \circ \varphi_\alpha, \tag{1.2.7}$$

and conversely, we can obtain any trivialization φ_α' over U_α in this manner via

$$f_\alpha := \varphi_\alpha' \circ \varphi_\alpha^{-1}$$

(φ_α^{-1} assigns to each x the diffeomorphism inverse to $\varphi_\alpha(x)$). $\tag{1.2.8}$

If f_α, f_β are as above, the transition maps change according to

$$\varphi_{\beta\alpha}^{-1} = \varphi_\beta' \circ \varphi_\alpha'^{-1} = f_\beta \circ \varphi_{\beta\alpha} \circ f_\alpha^{-1}. \tag{1.2.9}$$

The special case where all transition maps take their values in an *Abelian* subgroup A of Diff(F) yields some additional structure: The transition maps $\{\varphi_{\beta\alpha}\}$ then define a Čech cocycle on M with values in A, because (1.2.4) and (1.2.5) imply

$$\delta(\{\varphi_{\beta\alpha}\}) = 0$$

for the boundary operator δ. By (1.2.9), two such cocycles $\{\varphi_{\beta\alpha}\}$ and $\{\varphi'_{\beta\alpha}\}$ define the same bundle if $\{\varphi_{\beta\alpha}^{-1} \circ \varphi'_{\beta\alpha}\}$ is a coboundary. Thus, in this case, we can consider a bundle as a cohomology class in $H^1(M, A)$.[6]

A section of E is a smooth map

$$s : M \to E$$

satisfying

$$\pi \circ s = \text{id}.$$

We denote the space of sections by $C^\infty(E)$ or $\Gamma(E)$.

For our purposes, we shall only need two special (closely related) types of fiber bundles. The fiber F will be either a vector space V or a Lie group G. The important general principle here is to require that the transition maps respect the corresponding structure. Thus, they are not allowed to assume arbitrary values in Diff(F), but only in some fixed Lie group G. G is called the structure group of the bundle.

According to this principle, the fiber of a vector bundle is a real or complex vector space V of some real dimension n, and the structure group is $Gl(n, \mathbb{R})$ or some subgroup. A bundle whose fiber is a Lie group G is called a principal bundle, and the total space is denoted by P. The structure group is G or some subgroup, and it operates by left multiplication on the fiber G. Right multiplication on G induces a right action of G on P via local trivializations:

$$P \times G \to P, \qquad (x, g) * h = (x, gh) \quad \text{for } p = (x, g) \in P,$$

with the composition rule $(p * g)h = p * gh$. This action is free, that is, $p * g = p \Leftrightarrow g = e$ (neutral element). The projection $\pi : P \to M$ is obtained by simply identifying $x \in M$ with an orbit of this action, that is,

$$\pi : P \to P/G = M.$$

The groups $Gl(n, \mathbb{R})$, $O(n)$, $SO(n)$, $U(n)$ and $SU(n)$ will be the ones of interest for us. Acting as linear groups on a vector space, they preserve linear, Euclidean, or Hermitian structures. For example, a Euclidean structure, that is, a (positive definite) scalar product, is an additional structure on a vector space. According to the general principle, if we have such a structure on our fiber, it has to be respected by the transition maps. As before, this restricts the transformations permitted. In our example, we thus allow only $O(n)$ in place of $Gl(n, \mathbb{R})$. Such a restriction of the admissible transformations by imposing an additional structure that has to be preserved is called a reduction of the structure group.

---

[6] We assume here that M is connected; otherwise, in place of A itself, we should utilize the locally constant sheaf of A.

Principal and vector bundles are closely related. Let $P \to M$ be a principal bundle with fiber G, and let the vector space V carry a representation of G. We then construct a vector bundle E with fiber V using the following free right action of G on $P \times V$:

$$P \times V \times G \to P \times V,$$

$$(p, v) * g = (p * g, g^{-1}v).$$

The projection

$$P \times V \to P \to M$$

is invariant under this action, and

$$E := P \times_G V := (P \times V)/G \to M$$

is a vector bundle with fiber

$$G \times_G V = (G \times V)/G = V$$

and structure group G. Via the left action of G on V, the transition maps for P yield transition maps for E. Conversely, if we have a vector bundle with structure group G, we construct a G-principal bundle P by

$$\coprod_\alpha U_\alpha \times G/\sim$$

with

$$(x_\alpha, g_\alpha) \sim (x_\beta, g_\beta) :\Leftrightarrow x_\alpha = x_\beta \text{ in } U_\alpha \cap U_\beta \quad \text{and} \quad g_\beta = \varphi_{\beta\alpha}(x)g_\alpha.$$

(Here, $\{U_\alpha\}$ is a local trivialization of E with transition maps $\varphi_{\beta\alpha}$; these transition maps are in $Gl(n, \mathbb{R})$. Since the elements g_α are in the structure group G which is assumed to be a linear group, that is, a subgroup of $Gl(n, \mathbb{R})$, we can form the product $\varphi_{\beta\alpha}(x)g_\alpha$.) P can be viewed as the bundle of admissible bases of E. In a local trivialization, each fiber of E is identified with $\mathbb{R}^n$ or $\mathbb{C}^n$, and each admissible base is represented by a matrix with coefficients in $\mathbb{R}$ or $\mathbb{C}$. The transition maps then effect a base change. In each local trivialization, the action of G on P is given by matrix multiplication.

All standard operations on vector spaces extend to vector bundles. If we have a vector bundle E with fiber V_x over x, we can form the dual bundle E^* with fiber the dual space $V_x^\star$ of V_x. Applying this construction to the tangent bundle TM yields the cotangent bundle T^*M. If E_1 and E_2 are vector bundles, we can form the bundles $E_1 \oplus E_2$, $E_1 \otimes E_2$ and $E_1 \wedge E_2$ by performing the corresponding operations on the fibers. In particular, from the cotangent bundle T^*M, we obtain the bundle $\Lambda^p(M)$ introduced in (1.1.23), whose sections are the exterior p-forms.

1.2.2 Covariant Derivatives

Let E be a vector bundle over M. We may view E as a family of vector spaces parametrized by M. A local trivialization φ over U identifies the fibers over U with each other. Changing the local trivialization then also changes this identification of the fibers. The identification thus depends on the choice of a local trivialization and is therefore not canonical. Hence, while we can decide whether a section of E is differentiable because all transition maps depend differentiably on x and therefore do not affect the differentiability of a section in some local trivialization, there is no canonical way to specify the value of its derivative. In particular, we do not have a criterion for a section being constant along a curve in M.

Therefore, in order to be able to differentiate a section, we need to introduce and specify an additional structure on E, a so-called covariant derivative or connection. We point out that this includes and generalizes the concept of a covariant derivative developed in Sect. 1.1.1 for the tangent bundle.

A covariant derivative is an operator

$$D : \Gamma(E) \to \Gamma(E) \otimes_{C^\infty(M)} \Gamma(T^*M)$$

with the following properties: For $\sigma \in \Gamma(E)$, $V \in T_xM$, we write

$$D\sigma(V) =: D_V\sigma$$

and require (for all $x \in M$):

(i) D is tensorial in V:

$$D_{V+W}\sigma = D_V\sigma + D_W\sigma \quad \forall V, W \in T_xM, \ \sigma \in \Gamma(E),$$

$$D_{fV}\sigma = f D_V\sigma \quad \forall V \in T_xM, \ f \in C^\infty(M), \ \sigma \in \Gamma(E).$$

(Remark: It does not really make sense to multiply an element $V \in T_xM$ by a function $f \in C^\infty(M)$. What the preceding rule means is that when we take a section $V \in \Gamma(TM)$ of the tangent bundle, the value $(D_V\sigma)(x)$ depends only on the value of V at the point x, but not on the values at other points. This is not so for σ as rule (iii) shows.)

(ii) D is linear in σ:

$$D_V(\sigma + \tau) = D_V\sigma + D_V\tau \quad \forall V \in T_xM, \ \sigma, \tau \in \Gamma(E).$$

(iii) D satisfies the product rule:

$$D_V(f\sigma) = V(f)\sigma + f D_V\sigma \quad \forall V \in T_xM, \ f \in C^\infty(M), \ \sigma \in \Gamma(E).$$

An example, which is not really typical, but in a certain sense a local model, is the trivial bundle $M \times \mathbb{R}$ over M, where we can put

$$D_V\sigma := d\sigma(V) = V(\sigma)$$

to obtain a covariant derivative. In the general case, let φ be a trivialization of E over U,

$$E_{|U} \cong U \times \mathbb{R}^n \quad (=\varphi(\pi^{-1}(U))).$$

Via this local identification, a base of $\mathbb{R}^n$ yields a base $\mu_1, \ldots, \mu_n$ of sections of $E_{|U}$. Any section σ can then be written over U as

$$\sigma(x) = a^k(x)\mu_k(x).$$

Then

$$D\sigma = (da^k)\mu_k + a^k D\mu_k. \tag{1.2.10}$$

Since $(\mu_j)_{j=1,\ldots,n}$ is a base of sections, we can write

$$D\mu_k = A_k^j \mu_j. \tag{1.2.11}$$

Here, for each x, $A(x) = (A_k^j(x))_{j,k=1,\ldots,n}$ is a $T_x^* M$-valued matrix, that is, an element of $\mathfrak{gl}(n, \mathbb{R}) \otimes T_x^* M$. In symbols,

$$A \in \Gamma\big(\mathfrak{gl}(n, \mathbb{R}) \otimes T^* M_{|U}\big).$$

In our trivialization, we write this as

$$D = d + A. \tag{1.2.12}$$

We now wish to determine the transformation behavior of A under a change of the local trivialization. Let $\{U_\alpha\}$ be an open covering of M that yields a local trivialization with transition maps

$$\varphi_{\beta\alpha} : U_\alpha \cap U_\beta \to Gl(n, \mathbb{R}).$$

D defines a $T^* M$-valued matrix A_α on U_α. Let the section μ be represented by μ_α on U_α. A Greek index α here is not a coordinate index, but refers to the chosen covering $\{U_\alpha\}$. Thus

$$\mu_\beta = \varphi_{\beta\alpha}\mu_\alpha \quad \text{on } U_\alpha \cap U_\beta.$$

This implies

$$\varphi_{\beta\alpha}(d + A_\alpha)\mu_\alpha = (d + A_\beta)\mu_\beta \quad \text{on } U_\alpha \cap U_\beta. \tag{1.2.13}$$

On the left-hand side, we have first computed $D\mu$ in the local trivialization determined by U_α and then transformed the result into the local trivialization determined by U_β, while on the right-hand side, we have directly expressed $D\mu$ in the latter. We conclude

$$A_\alpha = \varphi_{\beta\alpha}^{-1} d\varphi_{\beta\alpha} + \varphi_{\beta\alpha}^{-1} A_\beta \varphi_{\beta\alpha}. \tag{1.2.14}$$

We have thus found the transformation behavior. A_α does not transform as a tensor, because of the term $\varphi_{\beta\alpha}^{-1} d\varphi_{\beta\alpha}$. The difference of two connections, however, does transform as a tensor. The space of all connections on a given vector bundle is therefore an affine space. The difference of two connections is a $\mathfrak{gl}(n, \mathbb{R})$-valued 1-form.

Having a connection D on a vector bundle E, it is now our aim to extend D to associated bundles, requiring suitable compatibility conditions. We start with the dual bundle E^*. Let

$$(\cdot, \cdot) : E \otimes E^* \to \mathbb{R}$$

be the bilinear pairing between E and E^*. The base dual to some base $\mu_1, \ldots, \mu_n$ of E is denoted by $\mu^{*1}, \ldots, \mu^{*n}$, i.e.,

$$(\mu_i, \mu^{*j}) = \delta_i^j. \qquad (1.2.15)$$

We then define the connection D^* on E^* by requiring

$$d(\mu, v^*) = (D\mu, v) + (\mu, D^*v) \qquad (1.2.16)$$

for all $\mu \in \Gamma(E), v \in \Gamma(E^*)$. In our above notation

$$D = d + A, \qquad D^* = d + A^*. \qquad (1.2.17)$$

From (1.2.15) (cf. (1.2.11)) we then compute

$$0 = d(\mu_i, \hat{\mu}^j) = (A_i^k \mu_k, \mu^{*j}) + (\mu_i, A_i^{*j} \mu^{*l})$$
$$= A_i^j + A_i^{*j},$$

i.e.,

$$A^* = -A. \qquad (1.2.18)$$

We now construct a connection on a product bundle from connections on the factors. If E_1 and E_2 are vector bundles over M with connections D_1, D_2, we obtain a connection D on $E := E_1 \otimes E_2$ by

$$D(\mu_1 \otimes \mu_2) = D_1\mu_1 \otimes \mu_2 + \mu_1 \otimes D_2\mu_2$$
$$(\mu_i \in \Gamma(E_i), \; i = 1, 2). \qquad (1.2.19)$$

We apply this construction to $\text{End}(E) = E \otimes E^*$ to obtain a connection that is again denoted by D. For a section $\sigma = \sigma_j^i \mu_i \otimes \mu^{*j}$, we then have

$$D(\sigma_j^i \mu_i \otimes \mu^{*j}) = d\sigma_j^i \mu_i \otimes \mu^{*j} + \sigma_j^i A_i^k \mu_k \otimes \mu^{*j} - \sigma_j^i A_k^j \mu_i \otimes \mu^{*k}$$
$$= d\sigma + [A, \sigma]. \qquad (1.2.20)$$

Thus, the connection induced on $\text{End}(E)$ operates via the Lie bracket. In a slightly different interpretation, we can view a connection D as a map

$$D : \Gamma(E) \to \Gamma(E) \otimes \Omega^1(M).$$

Using the notation

$$\Omega^p(E) := \Gamma(E) \otimes \Omega^p(M),$$

we extend D to a map

$$D : \Omega^p(E) \to \Omega^{p+1}(E)$$

by

$$D(\mu\omega) = D\mu \wedge \omega + \mu d\omega \qquad (1.2.21)$$

(where $\mu \in \Gamma(E)$, $\omega \in \Omega^p(M)$, and we have written $\mu\omega$ in place of $\mu \otimes_{C^\infty(M)} \omega$.[7])
The **curvature** of a connection D is now defined as

$$F := D^2 : \Omega^0(E) \to \Omega^2(E).$$

D is called flat if

$$F = 0.$$

Since the exterior derivative d satisfies (1.1.37), i.e.,

$$d \circ d = 0$$

we obtain the de Rham complex

$$\Omega^0 \xrightarrow{d} \Omega^1 \xrightarrow{d} \Omega^2 \xrightarrow{d} \ldots \quad (\Omega^p = \Omega^p(M)).$$

The sequence

$$\Omega^0(E) \xrightarrow{D} \Omega^1(E) \xrightarrow{D} \Omega^2(E) \xrightarrow{D} \ldots$$

however, is not necessarily a complex, since in general $F \neq 0$. For $\mu \in \Gamma(E)$ $(= \Omega^0(E))$, we compute

$$F(\mu) = (d + A) \circ (d + A)\mu$$
$$= (d + A)(d\mu + A\mu)$$
$$= (dA)\mu - A\, d\mu + A\, d\mu + A \wedge A\mu$$

(the minus sign arises because A takes values in 1-forms), that is,

$$F = dA + A \wedge A. \tag{1.2.22}$$

If we write

$$A = A_j\, dx^j$$

in local coordinates, with $A_j \in \Gamma(\mathfrak{gl}(n, \mathbb{R})) = \Gamma(\text{End}(E))$, (1.2.22) becomes

$$F = \frac{1}{2}\left(\frac{\partial A_j}{\partial x^i} - \frac{\partial A_i}{\partial x^j} + [A_i, A_j] \right) dx^i \wedge dx^j. \tag{1.2.23}$$

F is a map from $\Omega^0(E)$ to $\Omega^2(E)$, i.e.,

$$F \in \Omega^2(E) \otimes (\Omega^0(E))^* = \Omega^2(\text{End}(E)).$$

Therefore, according to our rules (1.2.20) and (1.2.21),

$$DF = dF + [A, F]$$
$$= dA \wedge A - A \wedge dA + [A, dA + A \wedge A] \quad \text{(by (1.2.22))}$$
$$= dA \wedge A - A \wedge dA + A \wedge dA - dA \wedge A + [A, A \wedge A]$$

---

[7] We leave it to the reader to (easily) verify that (1.2.21) is well defined, even though the decomposition $\mu \otimes_{C^\infty(M)} \omega$ is not canonical.

$$= [A, A \wedge A]$$
$$= [A_i \, dx^i, A_j \, dx^j \wedge A_k \, dx^k]$$
$$= A_i A_j A_k (dx^i \wedge dx^j \wedge dx^k - dx^j \wedge dx^k \wedge dx^i$$
$$\quad - dx^i \wedge dx^k \wedge dx^j + dx^k \wedge dx^j \wedge dx^i)$$
$$= 0.$$

We thus obtain the Bianchi identity:

Theorem 1.1 *The curvature of a connection D satisfies*

$$DF = 0. \tag{1.2.24}$$

The Bianchi identity can also be derived in a conceptually more interesting manner from the equivariance of the curvature ($f^* F_D = F_{f^*D}$, F_D = curvature of D) under bundle automorphisms f, that is, diffeomorphisms commuting with the group action (cf. [95]).

Using the notation of (1.2.13), we now wish to determine the transformation behavior of the curvature F of a connection D. From (1.2.14),

$$F_\alpha = dA_\alpha + A_\alpha \wedge A_\alpha$$
$$= d(\varphi_{\beta\alpha}^{-1} d\varphi_{\beta\alpha}) + d(\varphi_{\beta\alpha}^{-1} A_\beta \varphi_{\beta\alpha})$$
$$\quad + \varphi_{\beta\alpha}^{-1} d\varphi_{\beta\alpha} \wedge \varphi_{\beta\alpha}^{-1} d\varphi_{\beta\alpha} + \varphi_{\beta\alpha}^{-1} d\varphi_{\beta\alpha} \wedge \varphi_{\beta\alpha}^{-1} A_\beta \varphi_{\beta\alpha}$$
$$\quad + \varphi_{\beta\alpha}^{-1} A_\beta \wedge d\varphi_{\beta\alpha} + \varphi_{\beta\alpha}^{-1} A_\beta \wedge A_\beta \varphi_{\beta\alpha}.$$

Because of

$$d(\varphi_{\beta\alpha}^{-1}) = -\varphi_{\beta\alpha}^{-1} d\varphi_{\beta\alpha} \varphi_{\beta\alpha}^{-1}$$

the derivatives of $\varphi_{\beta\alpha}$ cancel, and we obtain

$$F_\alpha = \varphi_{\beta\alpha}^{-1} F_\beta \varphi_{\beta\alpha}. \tag{1.2.25}$$

Thus, F transforms as a tensor, in contrast to A.

1.2.3 Reduction of the Structure Group. The Yang–Mills Functional

We now wish to implement the general principle formulated above that additional structures on the fibers of a bundle lead to restrictions on the admissible transformations. In the previous section, $Gl(n, \mathbb{R})$ was the structure group of our vector bundle. This reflected the fact that we only had a linear (vector space) structure on our fibers, but nothing else. We shall now consider vector spaces with a structure group $G \subset Gl(n, \mathbb{R})$. The group G will then be interpreted as the invariance group of some structure on the fibers. Let $\mathfrak{g}$ be the Lie algebra of G. For a connection D on the vector bundle E with fiber $\mathbb{R}^n$, we then require compatibility with the G-structure. To make this more precise, we consider local trivializations

$$\varphi : \pi^{-1}(U) \to U \times \mathbb{R}^n$$

of E whose transition functions preserve the G-structure, that is, ones that transform G-bases $\mu_1, \ldots, \mu_n$ (meaning that the matrix with the columns $\mu_1, \ldots, \mu_n$ is contained in G) into G-bases. Linear algebra (Gram-Schmidt) tells us that we can always construct such trivializations. In such a trivialization, we also require of

$$D = d + A$$

that

$$A \in \Gamma(\mathfrak{g} \otimes T^* M_{|U}). \tag{1.2.26}$$

Let us consider some examples. $G = O(n)$ means that each fiber of E possesses a Euclidean scalar product $\langle \cdot, \cdot \rangle$. Via a corresponding local trivialization, for each $x \in U$, we then obtain an orthonormal base $e_1(x), \ldots, e_n(x)$ of the fiber V_x over x depending smoothly on x, namely $\varphi^{-1}(x, e_1, \ldots, e_n)$, where $e_1, \ldots, e_n$ is an orthonormal base of $\mathbb{R}^n$ w.r.t. the standard Euclidean scalar product. We then want that the Leibniz rule holds, i.e.,

$$d \langle \sigma, \tau \rangle = \langle D\sigma, \tau \rangle + \langle \sigma, D\tau \rangle, \tag{1.2.27}$$

that is, we require that $\langle \cdot, \cdot \rangle$ is covariantly constant. This implies in particular

$$0 = d \langle e_i, e_j \rangle = \langle A \, e_i, e_j \rangle + \langle e_i, A \, e_j \rangle, \tag{1.2.28}$$

that is, A is skew symmetric, $A \in \mathfrak{o}(n)$. A connection D satisfying the Leibniz rule is called a *metric connection*.

Analogously, for $G = U(n)$ we have a Hermitian product on the fibers, and the corresponding Leibniz rule implies

$$A \in \mathfrak{u}(n). \tag{1.2.29}$$

We then speak of a *Hermitian connection*.

$\mathrm{Ad}E$ is defined to be the bundle with fibers $(\mathrm{Ad}E)_x \subset \mathrm{End}(V_x)$ consisting of those endomorphisms of V_x that are contained in G. $\mathrm{Ad}E = P \times_G \mathfrak{g}$, where P is the associated principal bundle G acts on $\mathfrak{g}$ by the adjoint representation. Analogously, $\mathrm{Aut}(E)$ is the bundle with fiber G, now considered as the automorphism group of V_x, that is,

$$\mathrm{Aut}(E) = P \times_G G,$$

where G acts by conjugation. (Thus, $\mathrm{Aut}(E)$ is not a principal bundle.) (The reason for this action is the compatibility with the action

$$P \times V \times G \to P \times V, \qquad (p, v) * g = (pg, g^{-1}v),$$

because with $(p, h) * g = (pg, g^{-1}hg)$, we obtain $g^{-1}hg(g^{-1}v) = g^{-1}(hv)$, since G acts on V from the left.) Sections of $\text{Aut}(E)$ are called gauge transformations, and the group of gauge transformations is called the gauge group.

A section $s \in \Gamma(\text{Aut}(E))$ operates on a connection D by

$$s^*D = s^{-1} \circ D \circ s, \tag{1.2.30}$$

hence, for $\mu \in \Gamma(E)$

$$s^*(D)\mu = s^{-1}D(s\mu), \tag{1.2.31}$$

and, with $D = d + A$

$$s^*(A) = s^{-1}ds + s^{-1}A\,s. \tag{1.2.32}$$

In our present notation, the transformation rule (1.2.25) for the curvature F of D becomes

$$s^*(F) = s^{-1} \circ F \circ s. \tag{1.2.33}$$

Here, we consider F as an element of $\Omega^2(\text{Ad}E) = \Gamma(\text{Ad}E \otimes \Lambda^2 T^*M)$, and s acts trivially on the factor $\Lambda^2 T^*M$. The induced product on the fibers $\text{Ad}E_x \otimes \Lambda^2 T_x^*M$ that comes from the bundle metric of E and the Riemannian metric of M will be denoted by $\langle . , . \rangle$.

Definition 1.5 Let M be a compact Riemannian manifold with metric g, E a vector bundle with a bundle metric over M, D a metric connection on E with curvature $F_D \in \Omega^2(\text{Ad}E)$. The *Yang–Mills functional* applied to D is

$$YM(D) := \int_M \langle F_D, F_D \rangle dvol_g. \tag{1.2.34}$$

We now recall from Sect. 1.2.2 that the space of all connections on E is an affine space; the difference of two connections is contained in $\Omega^1(\text{End}E)$. Therefore, the space of all metric connections on E is an affine space as well; the difference of two metric connections is contained in $\Omega^1(\text{Ad}E)$. For deriving the Euler–Lagrange equations for the Yang–Mills functional, the variations to consider are therefore

$$D + tB \quad \text{with } B \in \Omega^1(\text{Ad}E).$$

For $\sigma \in \Gamma(E) = \Omega^0(E)$,

$$F_{D+tB}(\sigma) = (D + tB)(D + tB)\sigma$$

$$= D^2\sigma + tD(B\sigma) + tB \wedge D\sigma + t^2(B \wedge B)\sigma$$

$$= (F_D + t(DB) + t^2(B \wedge B))\sigma, \tag{1.2.35}$$

as $D(B\sigma) = (DB)\sigma - B \wedge D\sigma$. Therefore,

$$\frac{d}{dt} YM(D + tB)_{|t=0} = \frac{d}{dt} \int \langle F_{D+tB}, F_{D+tB} \rangle_{|t=0}$$

$$= 2 \int \langle DB, F_D \rangle. \tag{1.2.36}$$

Using the definition of D^* (1.2.17), this becomes

$$\frac{d}{dt} YM(D + tB)_{|t=0} = 2 \int \langle B, D^* F_D \rangle.$$

This vanishes for all variations B if

$$D^* F_D \equiv 0. \tag{1.2.37}$$

Definition 1.6 A metric connection D on the vector bundle E with a bundle metric over the Riemannian manifold M satisfying (1.2.37) is called a *Yang–Mills connection*.

In tensor notation, $F_D = F_{ij} dx^i \wedge dx^j$, and we want to express (1.2.37) in local coordinates with the normalization $g_{ij}(x) = \delta_{ij}$. In such coordinates,

$$d^*(F_{ij} dx^i \wedge dx^j) = -\frac{\partial F_{ij}}{\partial x^i} dx^j,$$

and hence from (1.2.18)

$$D^* F_D = \left(-\frac{\partial F_{ij}}{\partial x^i} - [A_i, F_{ij}] \right) dx^j.$$

The Yang–Mills equation (1.2.37) in local coordinates thus reads

$$\frac{\partial F_{ij}}{\partial x^i} + [A_i, F_{ij}] = 0 \quad \text{for } j = 1, \dots, d. \tag{1.2.38}$$

In the preceding, we have defined the Yang–Mills functional for metric connections, i.e., ones with structure group $G = O(n)$. Obviously, the same construction works for other compact structure groups, in particular for $U(m)$ and $SU(m)$. Those groups operate on the fibers of complex vector bundles. For a complex vector bundle, one has the structure group $Gl(m, \mathbb{C})$, that is, those of complex linear maps, and a Hermitian structure then, as explained, is a reduction of the structure group to $U(m)$. We now consider complex vector bundles, as for them, we can define important cohomology classes from the curvature of a connection, the so-called Chern classes, as we shall now explain. Thus, E now is a *complex* vector bundle of Rank m, that is, with fiber $\mathbb{C}^m$, over the compact manifold M. D is a connection on E with curvature

$$F = D^2 : \Omega^0 \to \Omega^2(E). \tag{1.2.39}$$

F satisfies the transformation rule (1.2.25):

$$F_\alpha = \varphi_{\beta\alpha}^{-1} F_\beta \varphi_{\alpha\beta}. \tag{1.2.40}$$

Therefore, we can consider F as an element of $\mathrm{Ad}E$. Since E is a complex vector bundle with structure group $\mathfrak{gl}(m, \mathbb{C})$, $\mathrm{Ad}E = \mathrm{End}E = \mathrm{Hom}_{\mathbb{C}}(E, E)$. Thus,

$$F \in \Omega^2(\mathrm{Ad}E), \qquad (1.2.41)$$

that is, F is a 2-form with values in the endomorphisms of E. Therefore, $\frac{i}{2\pi}F$ (the factor is simply chosen for convenient normalization) has eigenvalues $\lambda_k, k = 1, \ldots, m$, which are 2-forms. We then define exterior forms $c_j(E) \in \Omega^{2j}(M)$, $j = 1, \ldots, m$, on M via

$$\sum_{j=0}^{m} c_j(E)t^j = \det\left(\frac{i}{2\pi}tF + \mathrm{Id}\right) = \prod_{k=1}^{m}(1 + \lambda_k t). \qquad (1.2.42)$$

From the Bianchi identity (1.2.24), i.e., $DF = 0$, one concludes that $dc_j(E) = 0$ for all j. Thus, the $c_j(E)$ are closed and therefore represent cohomology classes. One also verifies that these classes do not depend on the choice of the connection D on E. These cohomology classes are called the *Chern classes* of the complex vector bundle E over M. Thus, from an arbitrary Hermitian connection on the bundle E, we can compute topological invariants of E and M.

For $j = 1, 2$, we get

$$c_1(E) = \frac{i}{2\pi}\,\mathrm{tr}\,F, \qquad (1.2.43)$$

$$c_2(E) - \frac{m-1}{2m}c_1(E) \wedge c_1(E) = \frac{1}{8\pi^2}\,\mathrm{tr}(F_0 \wedge F_0), \qquad (1.2.44)$$

where

$$F_0 := F - \frac{1}{m}\,\mathrm{tr}\,F \cdot \mathrm{Id}_E \quad \text{is the trace-free part of } F. \qquad (1.2.45)$$

We now consider a $U(m)$ vector bundle E over a *four-dimensional* oriented Riemannian manifold M. We let D be a Hermitian connection on E with curvature $F = D^2$. As explained in (1.1.29), (1.1.31), we can decompose F_0 into its self-dual and antiself-dual components

$$F_0 = F_0^+ + F_0^-. \qquad (1.2.46)$$

We recall (1.1.30), i.e., that the exterior product of a self-dual 2-form with an antiself-dual one vanishes, and obtain

$$\mathrm{tr}(F_0 \wedge F_0) = \mathrm{tr}(F_0^+ \wedge F_0^+) + \mathrm{tr}(F_0^- \wedge F_0^-)$$
$$= -|F_0^+|^2 + |F_0^-|^2 \qquad (1.2.47)$$

by (1.1.29) (note that the trace is the negative of the Killing form of the Lie algebra $\mathfrak{u}(m)$, that is, $A \cdot B = -\mathrm{tr}(AB)$, which explains the difference in sign between (1.2.47) and (1.1.29)).

From (1.2.44), we obtain by integration over M

$$(c_2(E) - \frac{m-1}{2m}c_1(E)^2)[M] = -\frac{1}{8\pi^2}\int(|F_0^+|^2 - |F_0^-|^2)\sqrt{g}dx^1 \wedge \cdots \wedge dx^d.$$
$$(1.2.48)$$

The Yang–Mills functional then can be decomposed as

$$YM(D) = \int_M \left(\frac{1}{m}|\operatorname{tr} F|^2 + |F_0|^2\right)\sqrt{g}dx^1 \wedge \cdots \wedge dx^d$$

$$= \int_M \left(\frac{1}{m}|\operatorname{tr} F|^2 + |F_0^+|^2 + |F_0^-|^2\right)\sqrt{g}dx^1 \wedge \cdots \wedge dx^d. \qquad (1.2.49)$$

Since $\operatorname{tr} F$ represents the cohomology class $-2\pi i c_1(E)$, the cohomology class of $\operatorname{tr} F$ is fixed, and

$$\int_M |\operatorname{tr} F|^2 \sqrt{g}\, dx^1 \wedge \cdots \wedge dx^d$$

becomes minimal if $\operatorname{tr} F$ minimizes the L^2-norm in this class ($\operatorname{tr} F$ therefore has to be a harmonic 2-form, see (1.1.115), (1.1.117)). Next, because of the constraint (1.2.48), that is, because the difference of the two integrals of F_0^+ and F_0^- is fixed by the topology of M and E, and therefore, $\int |F_0|^2$ becomes minimal if one of them vanishes, i.e.,

$$F_0^+ = 0 \quad \text{or} \quad F_0^- = 0, \qquad (1.2.50)$$

i.e. if F_0 is antiself-dual or self-dual. Which of these two alternatives can hold depends on the sign of $(c_2(E) - \frac{m-1}{m}c_1(E)^2)[M]$.

We now assume that the structure group of the complex vector bundle E is reduced to $SU(m)$. Thus, the fiber of $\operatorname{Ad} E$ is $\mathfrak{su}(m)$ which is trace-free. Therefore, if D is an $SU(m)$ connection, its curvature $F \in \Omega^2(\operatorname{Ad} E)$ satisfies

$$\operatorname{tr} F = 0. \qquad (1.2.51)$$

Consequently, by (1.2.43)

$$c_1(E) = 0,$$

and by (1.2.44), (1.2.48)

$$c_2(E)[M] = -\frac{1}{8\pi^2} \int_M (|F^+|^2 - |F^-|^2)\sqrt{g}dx^1 \wedge \cdots \wedge dx^d.$$

Thus again, the difference of the two parts of the Yang–Mills functional is topologically fixed, and as in (1.2.49),

$$YM(D) = \int_M (|F^+|^2 + |F^-|^2)\sqrt{g}dx^1 \wedge \cdots \wedge dx^d$$

is therefore minimized if F is (anti)self-dual; again, which of the two possibilities can hold depends on the sign of $c_2(E)[M]$. We conclude that:

Theorem 1.2 *For a vector bundle E with structure group $SU(m)$ over a compact oriented four-dimensional manifold M, an $SU(m)$ connection D on E yields an* *absolute minimum for the Yang–Mills functional if its curvature F is self-dual or antiself-dual.*

For a systematic presentation of four-dimensional Yang–Mills theory, we refer to [31].

1.2.4 The Kaluza–Klein Construction

Here, we take up the discussion of Sect. 1.1.6 and combine it with a bundle construction. The idea was first put forward by Theodor Kaluza in order to unify gravity with electromagnetism. Although this was not successful in its original form, the general idea is still important and alive today.

Kaluza's ansatz was to consider, in place of the Lorentz manifold M, a fiber bundle $\bar{M}$ over M. Kaluza took the real axis $\mathbb{R}$ as the fiber. This was then modified by Oscar Klein who chose the fiber S^1, that is, the compact Abelian Lie group $U(1)$, and this is also what we shall do here. Subsequently, we shall consider more general fibers. Following the physics literature, we shall always assume that $\bar{M}$ is a principal fiber bundle.

We obtain a metric

$$\bar{g} = \pi^* g + \bar{A} \otimes \bar{A} \tag{1.2.52}$$

on $\bar{M}$ where $\pi : \bar{M} \to M$ is the projection, g is the Lorentz metric on M, and $\bar{A}$ is the 1-form for some $U(1)$ connection on $\bar{M}$. (More precisely, $\bar{A} = \pi^* A$ where A is the connection form on M.) As in Sect. 1.1.6, we take the total scalar curvature as our action functional, that is,

$$\mathcal{L}(\bar{g}) = \int_{\bar{M}} \bar{R} \sqrt{-\bar{g}} \, dx^0 \cdots dx^3 d\xi, \tag{1.2.53}$$

where $\bar{R}$ is the scalar curvature of $\bar{g}$ and ξ is the fiber coordinate. To rewrite this functional, we first give the formulae for the Ricci curvature of $\bar{g}$. Let $\bar{V}$ be a unit vector field in the fiber direction. Because of the form (1.2.52) of the metric, this simply means that $\bar{V}$ is dual to $\bar{A}$. For each tangent vector field X on M, we consider the horizontal lift $\bar{X}_h$ determined by

$$\pi_* \bar{X}_h = X \quad \text{and} \quad \bar{g}(\bar{X}_h, \bar{V}) = 0.$$

Let F be the curvature form for the connection A, i.e.,

$$F = dA \quad (\text{and } \pi^* F = d\bar{A})$$

(note that A is a $U(1)$ connection, hence Abelian, and so, here we do not have an $A \wedge A$ term in the formula for the curvature).

We then have, for the Ricci tensor $\bar{R}(\cdot, \cdot)$ of $\bar{g}$,

$$\bar{R}(\bar{X}_h, \bar{Y}_h) = R(X, Y) + 2F \circ F(X, Y), \tag{1.2.54}$$

where $R(\cdot, \cdot)$ is the Ricci tensor of M, and in local coordinates with

$$F = F_{\alpha\beta}\, dx^\alpha \wedge dx^\beta,$$

we have

$$(F \circ F)_{\alpha\beta} = g^{\gamma\delta} F_{\alpha\gamma} F_{\delta\beta} = -g^{\gamma\delta} F_{\alpha\gamma} F_{\beta\delta} \tag{1.2.55}$$

and

$$\bar{R}(\bar{X}_h, \bar{V}) = -d^* F(X), \tag{1.2.56}$$

$$\bar{R}(\bar{V}, \bar{V}) = |F|^2 = F^{\alpha\beta} F_{\alpha\beta}. \tag{1.2.57}$$

In particular, the scalar curvature satisfies

$$\bar{R} = \operatorname{tr} \bar{R}(\cdot, \cdot) = R - |F|^2, \tag{1.2.58}$$

where, of course, R is the scalar curvature of g. Upon integration over the fibers, (1.1.177) hence becomes

$$\mathcal{L}(\bar{g}) = \int_M (R - |F|^2)\sqrt{-g}\, dx, \tag{1.2.59}$$

that is, the sum of the Einstein–Hilbert functional of the base and the Yang–Mills functional of the fiber. If the Einstein field equations for the vacuum hold for such a metric on $\bar{M}$, then, by (1.1.162), $\bar{M}$ has to have vanishing Ricci curvature, and then by (1.2.57), F has to vanish. Since F is supposed to represent the electromagnetic field, this does not constitute a desirable physical consequence of this ansatz.[8]

We can extend this construction to principal fiber bundles $\pi : \bar{M} \to M$ with compact non-Abelian structure group G. For that purpose, let g' be a G-invariant metric on the fiber, which we can then extend to all of $T\bar{M}$ with the help of a connection (given by a 1-form A). Let g again be a metric on the base M. For tangent vectors $\bar{W}, \bar{Z}$ on $\bar{M}$, we put

$$\bar{g}(\bar{Z}, \bar{W}) = g(\pi_* \bar{Z}, \pi_* \bar{W}) + g'(Z, W)$$

(where Z, W denote the projections of $\bar{Z}, \bar{W}$ onto the fibers), obtaining a metric on $\bar{M}$. If $\bar{U}$ and $\bar{V}$ are tangential to a fiber, we obtain, with notation analogous to that above,

$$\bar{R}(\bar{X}_h, \bar{Y}_h) = R(X, Y) - 2g^{\gamma\delta} g'(F_{\alpha\gamma}, F_{\beta\delta}) X^\alpha Y^\beta$$

$$\left(X = X^\alpha \frac{\partial}{\partial x^\alpha}, Y = Y^\alpha \frac{\partial}{\partial x^\alpha} \right), \tag{1.2.60}$$

---

[8]In an alternative interpretation, one might consider $\bar{g}$ as consisting of g and A and interpret the Euler–Lagrange equations for (1.2.59) as coupled Einstein–Maxwell equations for the metric g and the potential A. In that case, the undesired consequence that F has to vanish does not follow, but then we have a coupling rather than a unification of gravity and electromagnetism.

$$\bar{R}(\bar{X}_h, \bar{V}) = -g'(d^* F(X), \bar{V}) \tag{1.2.61}$$

$$\bar{R}(\bar{U}, \bar{V}) = \bar{R}'(\bar{U}, \bar{V}) + \det(g^{\gamma\delta})^{\frac{1}{2}} \sum_{\alpha,\beta} g'(F_{\alpha\beta}, \bar{U}) g'(F_{\alpha\beta}, \bar{V}). \tag{1.2.62}$$

The action functional becomes

$$\mathcal{L}(\bar{g}) = \int_{\bar{M}} (R + R' - |F|^2) \, dvol_{\bar{M}}. \tag{1.2.63}$$

(Here, R is integrated on the base and the result is multiplied with the volume of the fiber, whereas R' can be integrated on any fiber, by G-invariance, and the result is multiplied with the volume of M—assuming M to be compact again.)

The Einstein field equations for the vacuum now no longer require the vanishing of F. $|F|^2$ has to be constant, however, when those equations hold, and base and fiber must have constant scalar curvature. In fact, taking the trace in (1.2.62) yields constant scalar curvature in the fiber direction when the field equations hold, and because the scalar curvature of the metric on the fiber bundle is constant, the scalar curvature in the fiber direction also has to be constant. Taking the trace in (1.2.60) then yields constant scalar curvature on the base.

1.3 Tensors and Spinors

1.3.1 Tensors

We have already encountered the tangent bundle TM of a manifold M; its dual bundle is the cotangent bundle T^*M. The fiber of the tangent bundle over $p \in M$ is the tangent space T_pM, and the fiber of the cotangent bundle is the cotangent space T_p^*M.

Definition 1.7 A p times *contravariant* and q times *covariant* tensor (field) on a differentiable manifold M is a section of

$$\underbrace{TM \otimes \cdots \otimes TM}_{p \text{ times}} \otimes \underbrace{T^*M \otimes \cdots \otimes T^*M}_{q \text{ times}}. \tag{1.3.1}$$

We recall that on a complex manifold, we have the decompositions

$$T^{\mathbb{C}}M = T'M \oplus T''M, \qquad T^{*\mathbb{C}}M = T^{*'}M \oplus T^{*''}M, \tag{1.3.2}$$

which are invariant under (holomorphic) coordinate changes, and the transformation rules (1.1.86),

$$dz^j = \frac{\partial z^j}{\partial w^l} dw^l, \qquad dz^{\bar{k}} = \frac{\partial z^{\bar{k}}}{\partial w^{\bar{m}}} dw^{\bar{m}} \tag{1.3.3}$$

when $z = z(w)$. We can therefore also speak of covariant tensors of type (r, s), meaning sections of

$$\underbrace{T^{*'}M \otimes \cdots \otimes T^{*'}M}_{r\ \text{times}} \otimes \underbrace{T^{*''}M \otimes \cdots \otimes T^{*''}M}_{s\ \text{times}}. \qquad (1.3.4)$$

(Contravariant tensors are defined analogously, with the tangent bundle in place of the cotangent bundle.)

For simplicity, we now consider the case of complex dimension 1, that is, of a Riemann surface, in order not to have to bother with too many indices. The reader will surely be able to transfer the subsequent considerations to the case of an arbitrary (finite) dimension. We return to the conceptualization of variations described in (1.1.22), (1.1.39), (1.1.41) and perform a variation

$$z \mapsto z + \epsilon f(z) =: z + \epsilon \delta z \qquad (1.3.5)$$

with a *holomorphic* f. We want to determine the induced variation $\delta \omega$ of an (r, s)-form, that is, of an object of the type

$$\Omega(z, \bar{z}) = \omega(z, \bar{z})(dz)^r (d\bar{z})^s. \qquad (1.3.6)$$

Here, r and s are called the conformal weights of ω. Analogously to (1.1.41), we obtain the induced variation

$$\delta_{f, \bar{f}} \Omega(z, \bar{z}) = (r(\partial_z f) + s(\partial_{\bar{z}} \bar{f}) + f \partial_z + \bar{f} \partial_{\bar{z}}) \Omega(z, \bar{z}). \qquad (1.3.7)$$

$r + s$ is called the scaling dimension, because for $z \mapsto \lambda z, \lambda \in \mathbb{R}$,

$$\Omega = \omega(z, \bar{z})(dz)^r (d\bar{z})^s \mapsto \lambda^{r+s} \omega(\lambda z, \lambda \bar{z})(dz)^r (d\bar{z})^s. \qquad (1.3.8)$$

$r - s$ is called the conformal spin, because for $z \mapsto e^{-i\vartheta} z$,

$$\Omega = \omega(z, \bar{z})(dz)^r (d\bar{z})^s \mapsto e^{-i(r-s)\vartheta} \omega(e^{-i\vartheta} z, e^{i\vartheta} \bar{z})(dz)^r (d\bar{z})^s. \qquad (1.3.9)$$

1.3.2 Clifford Algebras and Spinors

Let V be a vector space of dimension n over a field F, which we shall take to be $\mathbb{R}$ or $\mathbb{C}$ in the sequel, equipped with a quadratic form $Q : V \times V \to F$. We then form the Clifford algebra $Cl(Q)$ as the quotient of the tensor algebra $\bigoplus_{k \geq 0} \underbrace{V \otimes \cdots \otimes V}_{k\ \text{times}}$
of V by the two-sided ideal generated by all elements of the form

$$v \otimes v - Q(v, v). \qquad (1.3.10)$$

In other words, the product in the Clifford algebra is

$$\{v, w\} := vw + wv = 2Q(v, w). \qquad (1.3.11)$$

Let $e_1, \ldots, e_n$ be a basis of V. This basis then satisfies

$$e_i e_j + e_j e_i = 2Q(e_i, e_j). \qquad (1.3.12)$$

The dimension of $Cl(Q)$ is 2^n, a basis being given by

$$e_0 := 1, \ e_{\alpha_1} e_{\alpha_2} \cdots e_{\alpha_k}, \quad \text{with } 1 \le \alpha_1 < \cdots < \alpha_k \le n. \tag{1.3.13}$$

We define the degree of $e_{\alpha_1} \cdots e_{\alpha_k}$ to be k. The degree of e_0 is 0. We let $Cl^k(Q)$ be the vector space of elements of $Cl(Q)$ of degree k. We also let $Cl^{ev}(Q)$ and $Cl^{odd}(Q)$ denote the space of elements of even and odd degree, resp. We have

$$Cl^0(Q) = \mathbb{R} \text{ or } \mathbb{C}$$
$$Cl^1(Q) = V, \tag{1.3.14}$$

whereas

$$Cl^2(Q) =: \mathfrak{spin}(Q) \tag{1.3.15}$$

is a Lie algebra with bracket

$$[a, b] := ab - ba. \tag{1.3.16}$$

It acts on $Cl^1(Q) = V$ via

$$\tau(a)v := [a, v] = av - va. \tag{1.3.17}$$

(Using (1.3.11), one verifies that for $a \in Cl^2(V)$, $v \in Cl^1(V)$, we have $av - va \in Cl^1(V)$.)

The simply connected Lie group with Lie algebra $\mathfrak{spin}(Q)$ is then denoted by $Spin(Q)$ and called the spin group. According to the general theory of representations of Lie groups (see e.g. [45]), representations of $\mathfrak{spin}(Q)$ lift to ones of $Spin(Q)$.

Example

1. $Q = 0$: This yields the so-called Grassmann algebra with multiplication rule

$$\vartheta_i \vartheta_j + \vartheta_j \vartheta_i = 0,$$

for some basis $\vartheta_1, \dots, \vartheta_n$.
2. For $F = \mathbb{R}$, consider the quadratic form Q with

$$Q(e_i, e_i) = \begin{cases} 1 & \text{for } i = 1, \dots p, \\ -1 & \text{for } i = p+1, \dots, n, \end{cases} \qquad Q(e_i, e_j) = 0 \quad \text{for } i \ne j$$

for some basis $e_1, \dots, e_n$ of V. Putting $q := n - p$, we denote the corresponding Clifford algebra by

$$Cl(p, q).$$

$p = 0$ yields the Clifford algebra $Cl(0, n)$ usually considered in Riemannian geometry. Of course, for given n, the Clifford algebra $Cl(p, q)(p + q = n)$ depends on the choice of $p \in \{0, \dots, n\}$. This is no longer so for the complexification

$$Cl^{\mathbb{C}}(n) := Cl(p, q) \otimes_{\mathbb{R}} \mathbb{C} \quad (p + q = n).$$

In fact, we have

$$Cl^{\mathbb{C}}(m) \cong \mathbb{C}^{2^n \times 2^n} \quad \text{for } m = 2n,$$

$$Cl^{\mathbb{C}}(m) \cong \mathbb{C}^{2^n \times 2^n} \oplus \mathbb{C}^{2^n \times 2^n} \quad \text{for } m = 2n + 1.$$

We define the Pauli matrices

$$\sigma_0 = \begin{pmatrix} 1 & 0 \\ 0 & 1 \end{pmatrix}, \quad \sigma_1 = \begin{pmatrix} 0 & 1 \\ 1 & 0 \end{pmatrix}, \quad \sigma_2 = \begin{pmatrix} 0 & -i \\ i & 0 \end{pmatrix}, \quad \sigma_3 = \begin{pmatrix} 1 & 0 \\ 0 & -1 \end{pmatrix}.$$

$$(1.3.18)$$

They form a basis of the space of 2×2 Hermitian matrices. We have

$$\{\sigma_i, \sigma_j\} := \sigma_i \sigma_j + \sigma_j \sigma_i = 2\delta_{ij}\sigma_0 \quad \text{for } i, j = 1, 2, 3. \qquad (1.3.19)$$

(Note the $+$ sign here: $\{\sigma_i, \sigma_j\}$ is an anticommutator, not a commutator.)
    The correspondence

$$e_0 \mapsto \sigma_0, \quad e_1 \mapsto \sigma_1, \quad e_2 \mapsto \sigma_3, \quad e_1 e_2 \mapsto -i\sigma_2$$

thus yields a two-dimensional representation of $Cl(2, 0)$, whereas mapping

$$e_1 \mapsto \sigma_1, \quad e_2 \mapsto i\sigma_2, \quad e_1 e_2 \mapsto -\sigma_3$$

yields one of $Cl(1, 1)$ and

$$e_1 \mapsto i\sigma_1, \quad e_2 \mapsto i\sigma_2, \quad e_1 e_2 \mapsto -i\sigma_3$$

yields one of $Cl(0, 2)$. The representations of $Cl(2, 0)$ and $Cl(1, 1)$ are both isomorphic to the algebra of real 2×2 matrices, whereas that of $Cl(0, 2)$ is isomorphic to the quaternions $\mathbb{H}$. In particular, for later reference, we emphasize that we have displayed here real representations of $Cl(2, 0)$ and $Cl(1, 1)$.
    Looking at $Cl(2, 0)$, which will be of particular interest for us, and extending the representation to the complexification, we make the following observation which we will subsequently place in a general context. $ie_1 e_2$ is represented by σ_2, and it anticommutes with both e_1 and e_2. Therefore, the representation of $Cl^2(2, 0) = \mathfrak{spin}(2, 0)$ leaves the eigenspaces of $ie_1 e_2$ invariant. In contrast, e_1 and e_2, that is, the elements of $Cl^1(2, 0)$, interchange them. (In particular, as a representation of $\mathfrak{spin}(2, 0)$, the representation is reducible; the two parts themselves are irreducible, however. Here, this is trivial, because they are one-dimensional, but the pattern is general.) The eigenvalues of $ie_1 e_2$ are ± 1, and its eigenspaces are generated in our representation by the vectors

$$\begin{pmatrix} 1 \\ i \end{pmatrix} \quad \text{and} \quad \begin{pmatrix} 1 \\ -i \end{pmatrix}.$$

The correspondence

$$e_0 \mapsto \sigma_0, \quad \ldots, \quad e_3 \mapsto \sigma_3$$

yields a two-dimensional representation of $Cl(3, 0)$.
    We define the Dirac matrices

$$\gamma^0 = \begin{pmatrix} \sigma_0 & 0 \\ 0 & -\sigma_0 \end{pmatrix}, \quad \gamma^j = \begin{pmatrix} 0 & \sigma_j \\ -\sigma_j & 0 \end{pmatrix}, \quad \text{for } j = 1, 2, 3,$$

$$\gamma^5 = i\gamma^0\gamma^1\gamma^2\gamma^3 = \begin{pmatrix} 0 & \sigma_0 \\ \sigma_0 & 0 \end{pmatrix},$$

where each 0 represents a 2×2 block; i.e., the γ^i are 4×4 matrices. The matrix γ^0 is Hermitian, while $\gamma^1, \gamma^2, \gamma^3$ are skew Hermitian. (This is expressed in the formula $\gamma^0\gamma^\mu\gamma^0 = \gamma^{\mu\dagger}$ for $\mu = 0, 1, 2, 3$.) They satisfy

$$\{\gamma^0, \gamma^0\} = 2I = \{\gamma^5, \gamma^5\},$$
$$\{\gamma^j, \gamma^j\} = -2I \quad \text{for } j = 1, 2, 3,$$
$$\{\gamma^i, \gamma^k\} = 0 \quad \text{for } i \neq k,$$

where I is the 4×4 identity matrix. Thus, we obtain a four-dimensional representation of $Cl(1, 3)$ and $Cl^{\mathbb{C}}(4)$, called the Dirac representation, by

$$e_i \mapsto \gamma^{i-1} \quad \text{for } i = 1, 2, 3, 4.$$

(Note: it might be better to denote the Dirac matrices by $\gamma^1, \ldots, \gamma^4$ instead of $\gamma^0, \ldots, \gamma^3$. Here, however we follow the convention in the physics literature; γ^5 will subsequently be denoted by Γ when we consider arbitrary dimensions.) We also consider

$$\sigma^{\mu\nu} = \frac{1}{2}[\gamma^\mu, \gamma^\nu],$$

where $[.,.]$ is an ordinary commutator. (Note: in the physics literature, there is an additional factor i in the definition of $\sigma^{\mu\nu}$.)

In the Dirac representation, we have

$$\sigma^{0i} = \begin{pmatrix} 0 & \sigma_i \\ \sigma_i & 0 \end{pmatrix}, \qquad \sigma^{ij} = -\sum_k \varepsilon_{ijk} i \begin{pmatrix} \sigma_k & 0 \\ 0 & \sigma_k \end{pmatrix}$$

$$\left(\varepsilon_{ijk} := \begin{cases} 1 & \text{if } (i, j, k) \text{ is an even permutation of } (1, 2, 3) \\ -1 & \text{if } (i, j, k) \text{ is an odd permutation of } (1, 2, 3) \\ 0 & \text{otherwise.} \end{cases} \right)$$

In the Weyl representation, we instead define

$$\gamma^0 = \begin{pmatrix} 0 & -\sigma_0 \\ -\sigma_0 & 0 \end{pmatrix}, \qquad \gamma^j = \begin{pmatrix} 0 & \sigma_j \\ -\sigma_j & 0 \end{pmatrix}, \qquad \text{for } j = 1, 2, 3,$$
$$\gamma^5 = i\gamma^0\gamma^1\gamma^2\gamma^3 = \begin{pmatrix} \sigma_0 & 0 \\ 0 & -\sigma_0 \end{pmatrix}.$$

In this case, we have

$$\sigma^{0i} = \begin{pmatrix} \sigma_i & 0 \\ 0 & -\sigma_i \end{pmatrix}, \qquad \sigma^{ij} = -\sum_k \varepsilon_{ijk} i \begin{pmatrix} \sigma_k & 0 \\ 0 & \sigma_k \end{pmatrix}.$$

Therefore, the action of the $\sigma^{\mu\nu}$ is reducible into two subspaces of (complex) dimension 2 each. Finally, we have the pseudo-Majorana representation, where all γ^μ

are purely imaginary:

$$\gamma^0 = \begin{pmatrix} 0 & \sigma_2 \\ \sigma_2 & 0 \end{pmatrix}, \qquad \gamma^1 = \begin{pmatrix} i\sigma_3 & 0 \\ 0 & i\sigma_3 \end{pmatrix},$$

$$\gamma^2 = \begin{pmatrix} 0 & -\sigma_2 \\ \sigma_2 & 0 \end{pmatrix}, \qquad \gamma^3 = \begin{pmatrix} -i\sigma_1 & 0 \\ 0 & -i\sigma_1 \end{pmatrix}.$$

We now wish to consider representations of $Cl(2n, 0)$ and $Cl^{\mathbb{C}}(2n)$ more abstractly. We consider the algebra generated by a basis $\gamma_1, \ldots, \gamma_{2n}$ of $\mathbb{R}^{2n}$ satisfying

$$\{\gamma_\mu, \gamma_\nu\} = 2\delta_{\mu\nu}$$

and set

$$a_1 := \frac{1}{2}(\gamma_1 + i\gamma_2), \qquad a_1^\dagger := \frac{1}{2}(\gamma_1 - i\gamma_2),$$

$$\vdots \qquad\qquad\qquad \vdots$$

$$a_n := \frac{1}{2}(\gamma_{2n-1} + i\gamma_{2n}), \qquad a_n^\dagger := \frac{1}{2}(\gamma_{2n-1} - i\gamma_{2n})$$

in $\mathbb{R}^{2n} \otimes \mathbb{C}$. In the physics literature, the $a_i, a_i^\dagger$ are called fermion annihilation and creation operators. We equip $\mathbb{C}^n$ with the coordinates $z^1 = x^1 + ix^2, \ldots, z^n = x^{2n-1} + ix^{2n}$. We let $\Lambda^{(0,q)}\mathbb{C}^n$ be the space of $(0, q)$-forms, i.e. the vector space of differential forms generated by

$$dz^{\bar{i}_1} \wedge \cdots \wedge dz^{\bar{i}_q}, \quad 1 \le i_1 < \cdots < i_q \le n \quad (dz^{\bar{1}} = dx^1 - idx^2, \text{ etc.})$$

We let $\varepsilon(dz^{\bar{j}})$ denote the exterior multiplication by $dz^{\bar{j}}$ from the left, i.e.,

$$\varepsilon(dz^{\bar{j}})(dz^{\bar{j}_1} \wedge \cdots \wedge dz^{\bar{j}_q}) = dz^{\bar{j}} \wedge dz^{\bar{j}_1} \wedge \cdots \wedge dz^{\bar{j}_q},$$

sending $(0, q)$-forms to $(0, q + 1)$-forms. Likewise, we let $\iota(dz^{\bar{j}})$ be the adjoint of $\varepsilon(dz^{\bar{j}})$ w.r.t. the natural metric on $\mathbb{C}^n$; thus

$$\iota(dz^{\bar{j}})(dz^{\bar{j}_1} \wedge \cdots \wedge dz^{\bar{j}_q})$$

$$= \begin{cases} 0 & \text{if } j \notin \{j_1, \ldots, j_q\}, \\ (-1)^{\mu-1} dz^{\bar{j}_1} \wedge \cdots \wedge \widehat{dz^{\bar{j}_\mu}} \wedge \cdots \wedge dz^{\bar{j}_q}) & \text{if } j = j_\mu. \end{cases}$$

We then obtain the desired representation by

$$a_j^\dagger \mapsto \varepsilon(dz^{\bar{j}}),$$

$$a_k \mapsto \iota(dz^{\bar{k}}).$$

Of course, one verifies that the formulae

$$\{a_i, a_j\} = 0, \qquad \{a_i^\dagger, a_j^\dagger\} = 0, \qquad \{a_i, a_j^\dagger\} = \delta_{ij}$$

are represented by

$$\{\varepsilon(dz^{\bar{j}}), \varepsilon(dz^{\bar{k}})\} = 0, \qquad \{\iota(dz^{\bar{j}}), \iota(dz^{\bar{k}})\} = 0, \qquad \{\varepsilon(dz^{\bar{j}}), \iota(dz^{\bar{k}})\} = \delta_{jk}.$$

The space

$$S := \Lambda^{(0,\cdot)}\mathbb{C}^n := \bigoplus_{q=0}^{n} \Lambda^{(0,q)}\mathbb{C}^n$$

on which $Cl^{\mathbb{C}}(2n)$ thus acts is called spinor space. The elements of S are called ((complex) Dirac) spinors. Since by (1.3.14), V is a subspace of its Clifford algebra, it therefore operates by multiplication on any representation of that Clifford algebra, in particular on S. This is called Clifford multiplication.

The representation S is not irreducible as a representation of $\mathfrak{spin}(2n, 0)$, however. To see this, we consider the "chirality operator"

$$\Gamma := i^{\frac{n}{2}}\gamma_1 \cdots \gamma_{2n}$$

(for $2n = 4$, one often writes γ_5 in place of Γ as explained above)

(with the usual exponential series, we can also write $\Gamma = \exp(i\pi N)$, with the "number operator" $N := \sum_{j=1}^{n} a_j^\dagger a_j$).

$$\{\Gamma, \gamma_\mu\} = 0 \quad \text{for } \mu = 1, \ldots, 2n,$$

$$\Gamma^2 = 1.$$

Thus, we may decompose $Cl^{\mathbb{C}}(n)$ into the eigenspaces $Cl^{\mathbb{C}}(n)^{\pm}$ of Γ for the eigenvalues ± 1, and these eigenspaces are interchanged by Clifford multiplication with any $v \in \mathbb{C}^n \backslash \{0\}$. Thus

$$P_{\pm} := \frac{1}{2}(1 \pm \Gamma)$$

project onto the eigenspaces of Γ, and we get a corresponding decomposition

$$S = S^+ \oplus S^-$$

into "positive and negative chirality spinors" (also called right- and left-handed spinors), or "Weyl spinors". If $p - q \equiv 0, 1, 2 \mod 8$, one may also find a real representation of $Cl(p, q)$. The corresponding spinors are called real or Majorana spinors. An important example is $n = 4, p = 3, q = 1$. Likewise, for $q - p \equiv 0, 1, 2 \mod 8$, there exist imaginary or pseudo-Majorana spinors.

The Lie algebra $\mathfrak{so}(n)$ consists of skew symmetric matrices. It is generated by the matrices M^{ij} with coefficients

$$(M^{ij})_{ab} = \delta_a^i \delta_b^j - \delta_a^j \delta_b^i.$$

They satisfy the commutation rules

$$[M^{ij}, M^{kl}] = -\delta^{ik} M^{jl} + \delta^{jk} M^{il} + \delta^{il} M^{jk} - \delta^{jl} M^{ik}.$$

These rules are also satisfied by

$$\sigma_{ij} := -\frac{1}{4}(\gamma_i \gamma_j - \gamma_j \gamma_i)$$

where $\gamma_1, \ldots, \gamma_n$ are a basis of $\mathbb{R}^n$ with $\{\gamma_i, \gamma_j\} = -\delta_{ij}$. (Note that this differs by a factor $-\frac{1}{2}$ from the convention employed in the definition of the Dirac and Weyl representations above.)

Thus

$$M^{ij} \mapsto \sigma_{ij}$$

yields a representation of $\mathfrak{so}(n)$ on $\mathbb{R}^n$; in fact, we may identify $\mathfrak{so}(n)$ with $\mathfrak{spin}(0, n)$. Since $\mathfrak{spin}(0, n) = Cl^2(0, n)$ we thus get an induced representation of $\mathfrak{so}(n)$ on the spinor space S. This representation, however, does not lift to one of $SO(n)$, but only to one of $Spin(0, n)$, the two-sheeted cover of $SO(n)$.

In the case when n is even, since each σ_{ij} is a sum of products of two γ_i, and since Clifford multiplication with each γ_i interchanges the eigenspaces of $S^{\pm}$ of Γ, σ_{ij} leaves these eigenspaces invariant.

To summarize: We have established an isomorphism $\mathfrak{so}(n) \longleftrightarrow \mathfrak{spin}(n)$. Thus, $\mathfrak{so}(n)$ operates on $Cl^1(0, n)$, and each representation of the Clifford algebra $Cl(0, n)$ therefore induces a representation of $\mathfrak{so}(n)$. In particular, in this manner, we obtain the spinor representation of $\mathfrak{so}(n)$ (which induces a double valued representation of $SO(n)$).

Remark The presentation here partly follows that of [22]. The original reference for Clifford modules is [6].

1.3.3 The Dirac Operator

As explained, since the vector space V is a subspace of the Clifford algebra $Cl(Q)$, it operates on any representation of that Clifford algebra. We can thus multiply a vector, an element of V, with a spinor, an element of S, that is, we have Clifford multiplication

$$V \times S \to S. \tag{1.3.20}$$

In fact, since multiplication by an element of V interchanges S^+ and S^-, we have an operation

$$V \times S^{\pm} \to S^{\mp}. \tag{1.3.21}$$

Denoting the representation by γ and letting $\frac{\partial}{\partial x^i}$ be the partial derivative in the direction of e_i, we can define the Dirac operator

$$\displaystyle{\not{D}} := \gamma(e_i)\frac{\partial}{\partial x^i}, \tag{1.3.22}$$

which operates on spinor fields. The square $\not{D} \circ \not{D}$ of the Dirac operator is then a linear combination of second derivatives; that linear combination depends on the quadratic form Q defining the Clifford algebra. If the quadratic form Q is represented by the identity matrix, that is, if we consider the Clifford algebra $Cl(n, 0)$, the square of the Dirac operator is the (negative definite) Laplace operator (see (1.1.103), (1.1.105))

$$\Delta = \sum_i \frac{\partial^2}{\partial x_i^2}. \tag{1.3.23}$$

In order to develop some structural insights, it is now useful to start with the complex case, or more precisely with a complex vector space V with a nondegenerate quadratic form Q. As Q is nondegenerate, it induces a nondegenerate bilinear form $\langle ., . \rangle$, w.r.t. which V is self-dual. Moreover, on a representation S of the Clifford algebra $Cl(Q)$, we can find a nondegenerate bilinear form $(., .)$ that is invariant under multiplication by $v \in V = Cl^1(Q)$:

$$(vs, t) = (s, vt) \tag{1.3.24}$$

for all $s, t \in S$. We can then use $(., .)$ to identify S with its dual S^*, and (1.3.20) then induces morphisms

$$\Gamma : S^* \times S^* \to V \tag{1.3.25}$$

and

$$\tilde{\Gamma} : S \times S \to V. \tag{1.3.26}$$

In (1.3.25), to any two elements of S^*, we assign a vector $v \in V$ that operates on a pair σ, τ of elements of S by $(v\sigma, \tau)$, cf. (1.3.20), (1.3.24).

Using bases $\{s_a\}$ and $\{e_\mu\}$ of S^* and V, we write (1.3.25) as

$$\Gamma(s_a, s_b) = \Gamma^\mu_{ab} e_\mu. \tag{1.3.27}$$

These morphisms are symmetric and equivariant w.r.t. the representation of $Cl(Q)$. Turning to the real case, the situation is not as convenient: we cannot always find real versions of these morphisms; they only exist in certain cases. This depends on the classification of Clifford algebras. They always exist for the Minkowski signature, that is, for the Clifford algebra $Cl(1, n-1)$, in any dimension n. They also exist for $Cl(2, 0)$, the case of particular interest for us.

1.3.4 The Lorentz Case

Let us also exhibit the relation between the orthogonal group and the spin group in the Lorentz case. There exist many references on this topic, including the classic [81]. Let $x = (x^0, x^1, x^2, x^3) \in \mathbb{R}^{1,3}$. We put

$$\langle x, x \rangle = x^0 x^0 - x^1 x^1 - x^2 x^2 - x^3 x^3. \tag{1.3.28}$$

The subgroup of $Gl(4, \mathbb{R})$ that preserves $\langle x, x \rangle$ is the Lorentz group $O(1, 3)$. It consists of two components that are distinguished by the value of the determinant, $+1$ or -1, and have otherwise the same properties. Thus, we consider the identity component $SO(1, 3)$ where the determinant is $+1$, without essential loss of generality.

We shall see that the corresponding spin group is $Sl(2, \mathbb{C})$, the group of complex 2×2-matrices with determinant 1. To x, we associate the Hermitian matrix

$$X := x^\mu \sigma_\mu = \begin{pmatrix} x^0 + x^3 & x^1 - ix^2 \\ x^1 + ix^2 & x^0 - x^3 \end{pmatrix}, \tag{1.3.29}$$

where $\sigma_0, \ldots, \sigma_3$ are the Pauli matrices. We first note that

$$\langle x, x \rangle = \det X. \tag{1.3.30}$$

Since we have

$$\mathrm{Tr}(\sigma_\mu \sigma_\nu) = 2\delta_{\mu\nu}, \tag{1.3.31}$$

we obtain

$$x^\mu = \frac{1}{2} \mathrm{Tr}(X\sigma_\mu) \tag{1.3.32}$$

as the inverse of the equation expressing X in terms of x.

In the physics literature, one writes the Hermitian matrix X as

$$\begin{pmatrix} X^{1\dot{1}} & X^{1\dot{2}} \\ X^{2\dot{1}} & X^{2\dot{2}} \end{pmatrix}. \tag{1.3.33}$$

By the Hermitian condition, $X^{\alpha\dot{\beta}} = \overline{X^{\beta\dot{\alpha}}}$ so that $X^{1\dot{1}}$ and $X^{2\dot{2}}$ are real. In this notation, (1.3.32) becomes

$$x^0 = \frac{1}{2}(X^{1\dot{1}} + X^{2\dot{2}}), \qquad x^1 = \frac{1}{2}(X^{1\dot{2}} + X^{2\dot{1}}),$$

$$x^2 = \frac{i}{2}(X^{1\dot{2}} - X^{2\dot{1}}), \qquad x^3 = \frac{1}{2}(X^{1\dot{1}} - X^{2\dot{2}}).$$

We may use the relation (1.3.29) between a vector x and a Hermitian matrix X to define an operation of $Sl(2, \mathbb{C})$ on $\mathbb{R}^{1,3}$ as follows:

For $A \in Sl(2, \mathbb{C})$, we put

$$T(A)X := X' := AXA^\dagger. \tag{1.3.34}$$

With indices, this is written as

$$X'^{\sigma\dot{\tau}} = A^\sigma_\beta \bar{A}^{\dot{\tau}}_{\dot{\gamma}} X^{\beta\dot{\gamma}}. \tag{1.3.35}$$

Here, the dotted indices refer to the transformation according to the conjugate complex of A, and this then explains the convention employed in (1.3.33). The fact that two As appear in (1.3.34) suggests that one consider this expression as a product: Instead of the 4-vector X, we take two spinors ϕ, χ that transform according to

$$\phi'^\alpha = A^\alpha_\beta \phi^\beta, \qquad \chi'^{\dot{\gamma}} = A^{\dot{\gamma}}_{\dot{\delta}} \chi^{\dot{\delta}}. \tag{1.3.36}$$

Their product then transforms like X in (1.3.35),

$$\phi'^\sigma \chi'^{\dot{\tau}} = A^\sigma_\beta \bar{A}^{\dot{\tau}}_{\dot{\gamma}} \phi^\beta \chi^{\dot{\gamma}}. \tag{1.3.37}$$

Using the above formulae, we can express (1.3.34) as a transformation of the vector x:

$$x'^\mu = \frac{1}{2}\operatorname{Tr}(X\sigma'_\mu) = \frac{1}{2}\operatorname{Tr}(AXA^\dagger\sigma_\mu) = \frac{1}{2}x^\nu\operatorname{Tr}(A\sigma_\nu A^\dagger\sigma_\mu). \tag{1.3.38}$$

Thus,

$$x' = Bx, \tag{1.3.39}$$

with

$$B^\mu_\nu = \frac{1}{2}\operatorname{Tr}(A\sigma_\nu A^\dagger\sigma_\mu) = \frac{1}{2}\operatorname{Tr}(\sigma_\mu A\sigma_\nu A^\dagger) \tag{1.3.40}$$

as the trace is invariant under cyclic permutations.

One may check from this that (1.3.34) induces a Lorentz transformation, but this can more easily be derived from the fact that (1.3.34) maps Hermitian matrices to Hermitian matrices and preserves the determinant (since $A \in Sl(2, \mathbb{C})$ has determinant 1), and (1.3.30) then implies that $\langle \cdot, \cdot \rangle = x^0x^0 - x^1x^1 - x^2x^2 - x^3x^3$ (see (1.3.28)) is preserved.

Also, this yields a homomorphism

$$T : Sl(2, \mathbb{C}) \to SO(1, 3)$$

with kernel $\{\pm 1\}$ ($\pm$ identity in $Sl(2, \mathbb{C})$ leads to the identity in $SO(1, 3)$ in (1.3.34)), and image the identity component of the Lorentz group.[9] $Sl(2, \mathbb{C})$ is the universal cover of the identity component of the Lorentz group, which is doubly connected. Therefore, in the physics literature, representations of $Sl(2, \mathbb{C})$ are usually considered as double-valued representations of $SO(1, 3)$.

We also observe that the homomorphism T in (1.3.34) maps $SU(2)$ to $SO(3)$. Namely we have, for $A \in SU(2)$,

$$\operatorname{Tr}(T(A)X) = \operatorname{Tr}(AXA^\dagger) = \operatorname{Tr}(AXA^{-1}) = \operatorname{Tr}(X) = 2x^0$$

in the notations of (1.3.29). Thus, $T(A)$ preserves x^0, and since $\langle x, x \rangle = x^0x^0 - x^1x^1 - x^2x^2 - x^3x^3$ is also preserved, it preserves

$$x^1x^1 + x^2x^2 + x^3x^3$$

and therefore yields an orthogonal transformation of the x^1, x^2, x^3 space. As before, this yields a twofold covering of $SO(3)$, and $SU(2) \cong Spin(3)$.

Since $SO(1, 3)$ acts by automorphisms on $\mathbb{R}^{1,3}$ which can be considered as a group of translations, we can form the semidirect product $SO(1, 3) \ltimes \mathbb{R}^{1,3}$ where $(B, b) \in SO(1, 3) \ltimes \mathbb{R}^{1,3}$ operates on $\mathbb{R}^{1,3}$ via $x \mapsto Bx + b$ and where "semidirect product" refers to the obvious composition rule. The group of all isometries of Minkowski space is the semidirect product $O(1, 3) \ltimes \mathbb{R}^{1,3}$, the Poincaré group, but it suffices for our purposes to consider its connected component containing the identity. Again, it is covered by $Sl(2, \mathbb{C}) \ltimes \mathbb{R}^{1,3}$.

---

[9]The Lorentz group has four connected components, all isomorphic to $SO(1, 3)$, and we obtain the other components by space- and time-like reflections from the identity component.

The irreducible unitary representations of $Sl(2, \mathbb{C}) \ltimes \mathbb{R}^{1,3}$ were classified by Wigner. We sketch here those aspects of the representation theory that are directly relevant for elementary particle physics. A mathematical treatment to which we refer for further details and which emphasizes the applications in physics is given in [98], whereas a comprehensive presentation from the perspective of physics can be found in [103]. Since $\mathbb{R}^{1,3}$ is a normal subgroup of $Sl(2, \mathbb{C}) \ltimes \mathbb{R}^{1,3}$, the study of the representations proceeds by describing the orbits of the action of $Sl(2, \mathbb{C})$ on $\mathbb{R}^{1,3}$, identifying the isotropy group of a point on each orbit, called the "little group" in physics, and then finding the representations of those isotropy groups. We know from (1.3.28) that

$$m^2 := \langle x, x \rangle = x^0 x^0 - x^1 x^1 - x^2 x^2 - x^3 x^3 \qquad (1.3.41)$$

is preserved by the action of $Sl(2, \mathbb{C})$ on $\mathbb{R}^{1,3}$. In particular, each orbit must be contained in a level set of m^2. Physically, m is the mass of the particle defined by the representation, and it then suffices to consider the case $m \geq 0$. Using the identification (1.3.29) of $x \in \mathbb{R}^{1,3}$ with the matrix

$$X := x^\mu \sigma_\mu = \begin{pmatrix} x^0 + x^3 & x^1 - ix^2 \\ x^1 + ix^2 & x^0 - x^3 \end{pmatrix},$$

for $m^2 > 0$, we can select the point

$$\begin{pmatrix} \pm m & 0 \\ 0 & \pm m \end{pmatrix},$$

depending on whether $x^0 > 0$ or < 0. Since this is a multiple of the identity matrix, its isotropy group, that is, the group of matrices leaving it invariant under conjugation, see (1.3.34), is $SU(2)$. As described for instance in [45, 75, 98], the irreducible unitary representations of $SU(2)$ come in a discrete family, parametrized by a half integer

$$L = 0, \frac{1}{2}, 1, \frac{3}{2}, \ldots \qquad (1.3.42)$$

which can be identified with the spin of the particle. Thus, the class of representations corresponding to an orbit with $m^2 > 0$ is described by the continuous parameter m^2 and the discrete parameter L from (1.3.42).

A point on an orbit with $m^2 = 0$ is

$$\begin{pmatrix} 2 & 0 \\ 0 & 0 \end{pmatrix},$$

and its isotropy group is defined by the invariance condition

$$A \begin{pmatrix} 2 & 0 \\ 0 & 0 \end{pmatrix} A^\dagger = \begin{pmatrix} 2 & 0 \\ 0 & 0 \end{pmatrix}.$$

This implies that A has to be of the form

$$A = \begin{pmatrix} e^{i\theta} & z \\ 0 & e^{i\theta} \end{pmatrix}$$

for some $z \in \mathbb{C}, \theta \in \mathbb{R}$. Looking at the conjugation

$$\begin{pmatrix} e^{i\theta} & 0 \\ 0 & e^{i\theta} \end{pmatrix} \begin{pmatrix} 1 & z \\ 0 & 1 \end{pmatrix} \begin{pmatrix} e^{i\theta} & 0 \\ 0 & e^{i\theta} \end{pmatrix} = \begin{pmatrix} 1 & ze^{2i\theta} \\ 0 & 1 \end{pmatrix},$$

we see that the isotropy group, denoted by $\tilde{E}(2)$, is a double cover (because of the angle 2θ) of the group of Euclidean motions $SO(2) \ltimes \mathbb{C}$. By the same strategy as before, for determining its representations, we should look at the orbits of the $SO(2)$ action on $\mathbb{C}$ which are the origin 0 and the concentric circles about 0. The representations corresponding to the latter do not occur in elementary particle physics. So, we are left with the origin whose isotropy group, the little group, is $SO(2)$. Its irreducible representations are all one-dimensional and labeled by

$$s = 0, \pm\frac{1}{2}, \pm 1, \dots \tag{1.3.43}$$

where the factor $\frac{1}{2}$ corresponds to the fact that the rotations were about an angle 2θ. The key for understanding the representations of $SU(2)$ is the following. The Lie algebra $\mathfrak{su}(2)$ is generated by

$$t_\mu := \frac{i}{2}\sigma_\mu, \quad \mu = 1, 2, 3 \tag{1.3.44}$$

with the Pauli matrices σ_μ, see (1.3.18). They satisfy

$$[t_\mu, t_\nu] = \epsilon_{\mu\nu\sigma} t_\sigma \tag{1.3.45}$$

with the totally antisymmetric tensor $\epsilon_{\mu\nu\sigma}$. The real matrices

$$e_+ := -i(t_1 - it_2) = \begin{pmatrix} 0 & 1 \\ 0 & 0 \end{pmatrix}, \quad e_+ := -i(t_1 + it_2) = \begin{pmatrix} 0 & 0 \\ 1 & 0 \end{pmatrix}, \tag{1.3.46}$$

$$h := -it_3 = \frac{1}{2}\begin{pmatrix} 1 & 0 \\ 0 & -1 \end{pmatrix}, \tag{1.3.47}$$

yield a basis of the Lie algebra $\mathfrak{sl}(2, \mathbb{R})$ and satisfy

$$[h, t_+] = t_+, \quad [h, t_-] = -t_-, \quad [t_+, t_-] = 2h. \tag{1.3.48}$$

From this, one deduces that when ρ is a representation of $\mathfrak{sl}(2, \mathbb{C})$ on a vector space V and v_λ is an eigenvector of $\rho(h)$ with eigenvalue λ, then $\rho(t_\pm)v_\lambda$ are eigenvectors of $\rho(h)$ with eigenvalues $\lambda \pm 1$. One then finds that the possible values of λ are $L, L-1, \dots, -L$ for some half integer $L = 0, \frac{1}{2}, 1, \dots$, see e.g. [45, 75, 98]. Since the eigenvalues are nondegenerate, the dimension of this representation is then $2L + 1$.

1.3.5 Left- and Right-handed Spinors

We now put the transformation rule (1.3.37) for the product of two spinors into a more general perspective that will be needed below in Sect. 2.2.1 for defining Lagrangians for spinors. According to our previous general discussion, in the present case of $\mathbb{R}^4 \cong \mathbb{C}^2$, the spinor space is a four-dimensional complex vector space, i.e., isomorphic to $\mathbb{C}^4$. We have already seen in Sect. 1.3.2 that the spinor representation is not irreducible as a representation of the spin group, but splits into the direct sum of two chiral representations, i.e., each spinor can be written as

$$\psi = \begin{pmatrix} \psi_L \\ \psi_R \end{pmatrix}. \tag{1.3.49}$$

ψ_L is called a left-handed, ψ_R a right-handed spinor.

$A \in Sl(2, \mathbb{C})$ then acts via

$$\psi_L \mapsto A\psi_L,$$
$$\psi_R \mapsto (A^\dagger)^{-1}\psi_R. \tag{1.3.50}$$

With $A = (A_k^j)_{j,k=1,2}$, we have

$$\psi_L^i \mapsto A_k^i \psi_L^k,$$
$$\psi_R^i \mapsto \tilde{A}_k^i \psi_R^k \quad \text{with } \tilde{A}_k^i \bar{A}_k^j = \delta_{ij}.$$

From (1.3.40), we also get with the help of (1.3.31)

$$B_\nu^\mu \sigma_\nu = A^\dagger \sigma_\mu A \tag{1.3.51}$$

(note that here the summation convention is used even though the position of the indices is not right—it would be better to write the σs with upper indices, but we refrain here from changing an established convention). Putting

$$S(A) = \begin{pmatrix} A & 0 \\ 0 & (A^\dagger)^{-1} \end{pmatrix}, \quad \psi = \begin{pmatrix} \psi_L \\ \psi_R \end{pmatrix}, \tag{1.3.52}$$

the action of A is described by

$$\psi \mapsto S(A)\psi. \tag{1.3.53}$$

In the Weyl representation, with

$$\gamma^0 = \begin{pmatrix} 0 & -\sigma_0 \\ -\sigma_0 & 0 \end{pmatrix},$$

we then have

$$S^{-1} = \gamma^0 S^\dagger \gamma^0. \tag{1.3.54}$$

Finally (1.3.51) implies

$$S^{-1}\gamma^\mu S = B_\nu^\mu \gamma^\nu. \tag{1.3.55}$$

For two left-handed spinors (see (1.3.49)) ϕ, χ,

$$\phi\chi := \varepsilon_{\alpha\beta}\phi^\alpha \chi^\beta \tag{1.3.56}$$

transforms as a scalar under the spinor representation; namely

$$\varepsilon_{\alpha\beta} A^\alpha_\gamma \phi^\gamma A^\beta_\delta \chi^\delta = A^1_\gamma A^2_\delta \phi^\gamma \chi^\delta - A^2_\gamma A^1_\delta \phi^\gamma \chi^\delta$$
$$= (A^1_1 A^2_2 - A^2_1 A^1_2)\phi^1 \chi^2 + (A^1_2 A^2_1 - A^2_2 A^1_1)\phi^2 \chi^1$$
$$= \det A (\phi^1 \chi^2 - \phi^2 \chi^1)$$
$$= \varepsilon_{\alpha\beta} \phi^\alpha \chi^\beta \quad (\text{since } \det A = 1 \text{ for } A \in Sl(2, \mathbb{C})).$$

Similarly

$$\phi^\alpha \sigma_{\mu,\alpha\dot\alpha} \bar\chi^{\dot\alpha} \tag{1.3.57}$$

transforms as a vector, for $\mu = 0, 1, 2, 3$. This can be better understood by considering full spinors

$$\psi = \begin{pmatrix} \psi_L \\ \psi_R \end{pmatrix}. \tag{1.3.58}$$

Following the physics notation, in the Lorentzian case, we define $\psi^\dagger$ as the complex conjugate of ψ, and the Dirac-conjugated spinor as

$$\bar\psi := \psi^\dagger \gamma^0. \tag{1.3.59}$$

In the Riemannian case, the γ^0 is omitted, that is,

$$\bar\psi := \psi^\dagger. \tag{1.3.60}$$

Thus, returning to the Lorentzian case, $\bar\psi \omega = \bar\psi_L \omega_R + \bar\psi_R \omega_L$. Then in the Weyl representation,

$$\bar\psi \gamma^\mu \omega \tag{1.3.61}$$

transforms as a vector. In fact, applying a transformation $A \in Sl(2, \mathbb{C})$, we get

$$\psi^\dagger S(A)^\dagger \gamma^0 \gamma^\mu S(A)\omega = \bar\psi \gamma^0 S(A)^\dagger \gamma^0 \gamma^\mu S(A)\omega$$
$$= \bar\psi S^{-1} \gamma^\mu S\omega \quad \text{by (1.3.54)}$$
$$= B^\mu_\nu \bar\psi \gamma^\nu \omega \quad \text{by (1.3.55),}$$

which is the required formula. Since $\bar\psi \gamma^\mu \omega = \psi^\dagger \gamma^0 \gamma^\mu \omega = \psi^\dagger_L \sigma_\mu \omega_L - \psi^\dagger_R \sigma_\mu \omega_R$, we also see why (1.3.57) transforms as a vector.

1.4 Riemann Surfaces and Moduli Spaces

1.4.1 The General Idea of Moduli Spaces

We start with some general principles; their meaning may become apparent only after reading the rest of this section, and the reader is advised to proceed when these principles are unclear and return to them later. It may be helpful, however, to try to understand the sequel in the light of these principles. In any case, the present section is more abstract and has more of a survey character than the preceding ones.

One is given a mathematical object with some varying structure. An example of such an object is a differentiable manifold S, and the structure could be a complex structure, a Riemannian metric—perhaps of a particular type—and so on. One wants to divide out all invariances; for example, one wants to identify all isometric metrics. The invariances usually constitute a (discrete or Lie) group. The resulting space of invariance classes is then a moduli space. This already suggests that there will be a problem (more precisely, singularities of the moduli space) caused by those particular instances of the structure that possess more invariances than the typical ones, for example, those Riemannian metrics that are highly symmetric. The reason is obvious, namely that for those instances, we need to divide out a larger group of invariances than for the other ones.

Heuristic guiding principle
The moduli space for structures of some given type carries a structure of the same type.

So, for example, we expect a moduli space of Riemannian metrics to carry a Riemannian metric itself, a moduli space of complex structures to be a complex space itself, a moduli space of algebraic varieties to be an algebraic variety itself.

Typically, the space of such structures is not compact, that is, these structures can degenerate. One then wishes to compactify the moduli space. The compactifying boundary then also contains (certain) degenerate versions of the structure. The choice of admissible degenerate structures—which need not be unique—can be subtle and should be carried out so that the resulting space is a Hausdorff space.

Often, one also wishes to get a fine moduli space M_{fine}. Let p be a point in the (ordinary, or coarse) moduli space M representing an instance g of a structure. M_{fine} then should be the fibration over M with the fiber over p being that g.

1.4.2 Riemann Surfaces and Their Moduli Spaces

A Riemann surface can be defined in several different ways, that is, through different types of structures. While these notions turn out to be equivalent in the end, they lead to different approaches to the moduli space of Riemann surfaces and equip that moduli space with different structures, according to the above principle. We shall now explain these different structures and also illustrate why they are interesting, in particular how they lead to different mathematical constructions and applications. For more details and proofs, we refer to [64] and other references cited subsequently. A profound knowledge of Riemann surface theory is useful for understanding conformal field theory and string theory mathematically. Let S be a compact differentiable orientable surface of genus p. If not explicitly stated otherwise, we assume $p > 1$.

The basic point is that one and the same such differentiable surface can carry a continuum of different Riemann surface structures. That is, there are many pairs Σ_1, Σ_2 of Riemann surfaces that are both diffeomorphic to S, but not equivalent as

Riemann surfaces. The moduli problem then consists of defining and understanding the space of all such Riemann surfaces (modulo holomorphic equivalence).

1. A Riemann surface Σ is a discrete (fixed point free, cocompact) faithful representation of the fundamental group $\pi_1(S)$ into $G := \mathrm{PSL}(2, \mathbb{R})$, determined up to conjugation by an element of G. The moduli space is the space of such representations modulo conjugation.

More precisely: A Riemann surface Σ is a quotient H/Γ, where $H = \{z = x + iy \in \mathbb{C} : y > 0\}$ is the Poincaré upper half plane and Γ is a discrete group of isometries with respect to the hyperbolic metric

$$\frac{1}{y^2} dz \wedge d\bar{z}. \tag{1.4.1}$$

Γ is a subgroup of the isometry group $\mathrm{PSL}(2, \mathbb{R})$ of H.[10] Here, $\mathrm{PSL}(2, \mathbb{R}) = \mathrm{SL}(2, \mathbb{R})/\pm 1$, acting on H via $z \mapsto \frac{az+b}{cz+d}$, with a, b, c, d satisfying $ad - bc = 1$ describing an element of $\mathrm{SL}(2, \mathbb{R})$. Γ should operate properly discontinuously and freely. It thus should not contain elliptic elements. This excludes singularities of the quotient H/Γ arising from fixed points of the action of Γ. In order to exclude cusps, that is, in order to ensure that H/Γ is compact, parabolic elements (see insertion below) of Γ also have to be excluded. Thus, all elements of Γ different from the identity should be hyperbolic.

Insertion: Here, a transformation $z \mapsto \frac{az+b}{cz+d}$ of H is called hyperbolic if it has two fixed points on the extended real axis $\bar{\mathbb{R}} = \partial H \cup \{\infty\}$, parabolic if it has one fixed point on $\bar{\mathbb{R}}$, and elliptic if it has a fixed point in H. Since the fixed points are computed to be $\frac{a-d}{2c} \pm \frac{1}{2c}\sqrt{(a+d)^2 - 4}$, the transformation is hyperbolic iff $|a + d| > 2$. The standard example of a hyperbolic transformation is $z \mapsto 2z$, with fixed points at 0 and ∞, and a parabolic one is given by $z \mapsto \frac{z}{z+1}$, which has its unique fixed point at 0. A hyperbolic transformation γ maps the hyperbolic geodesic l between its two fixed points p_1, p_2 (the semicircle through p_1 and p_2 orthogonal to the real axis) into itself, that is, it is a translation along the hyperbolic geodesic l. We can then easily visualize the operation of γ on H; it simply maps each geodesic orthogonal to l to another such geodesic orthogonal to l, with the shift already determined by the operation of γ on l. When we consider the example $z \mapsto 2z$, the invariant geodesic is the imaginary axis. The invariant geodesic in H becomes a closed geodesic on the surface H/Γ, with length given by the length of the shift. A parabolic transformation does not have a fixed geodesic, but instead rotates any geodesic through its fixed point into another such geodesic. Therefore, a parabolic transformation does not produce a closed geodesic in the quotient.

Γ is isomorphic to the fundamental group $\pi_1(S)$. Thus, a Riemann surface is described by a faithful representation ρ of $\pi_1(S)$ in $G := \mathrm{PSL}(2, \mathbb{R})$. This essentially leads to the approach of Ahlfors and Bers to Teichmüller theory. Here, we need to identify any two representations that only differ by a conjugation with an

---

[10]The isometries of H are the same as the conformal automorphisms of H, because of the conformal invariance of the metric.

element of G. Thus, we consider the space of faithful representations up to conjugacy. A representation can be defined by the images of the generators, that is, by $2p$ elements of G, and this induces a natural topology on the moduli space. In particular, this allows us to compute the dimension of the moduli space: Each of the $2p$ generator images is described by three real degrees of freedom (a, b, c, d satisfying the relation $ad - bc = 1$) which altogether yields $6p$ degrees of freedom. From this, we first need to subtract 3, the degrees of freedom for one generator, because the generators $a_1, b_1, \ldots, a_p, b_p$ of $\pi_1(\Sigma)$ are not independent, but satisfy the relation $a_1 b_1 a_1^{-1} b_1^{-1} \cdots a_p b_p a_p^{-1} b_p^{-1} = 1$. We also need to subtract another 3 to account for the freedom of conjugating by an element g of PSL$(2, \mathbb{R})$. Thus, the (real) dimension of the moduli space of representations of $\pi_1(\Sigma)$ in PSL$(2, \mathbb{R})$ modulo conjugations is $6p - 6$. This moduli space of representations of the fundamental group yields the Teichmüller space T_p. The moduli space M_p is a branched quotient of that space.

Singularities of the moduli space arise when the image Γ of ρ has more automorphisms than such a generic subgroup of G (whose only automorphisms are given by conjugations). Degenerations arise from limits of sequences of faithful, that is, injective representations ρ_n that are no longer injective. Just as the Riemann surfaces are obtained as quotients H/Γ, the moduli space M_p itself is likewise a quotient T_p/C of the Teichmüller space T_p by a discrete group, the so-called mapping class group. (This Teichmüller space T_p is a complex space diffeomorphic—but not biholomorphic—to $\mathbb{C}^{3p-3}$. The complex structure was described by Bers through a holomorphic embedding into some complex Banach space. For recent results about this complex structure, we refer to [14]. T_p parametrizes marked Riemann surfaces, that is, Riemann surfaces together with a choice of generators of the first homology group. Since all automorphisms of a hyperbolic Riemann surface act nontrivially on the first homology, Teichmüller space does not suffer from the problem of the moduli space, that Riemann surfaces with nontrivial automorphism groups can create singularities.)

This approach is also useful because it can be generalized to moduli spaces of representations of the fundamental group of a Kähler manifold in some linear algebraic group G. This is called non-Abelian Hodge theory and leads to profound insights into the structure of Kähler manifolds. In particular, because such representations can be shown to factor through holomorphic maps, this leads to the at present strongest approach to a general structure theory of Kähler manifolds via the Shafarevitch conjecture, see, e.g., [70–72].

2. A Riemann surface Σ is a 1-dimensional complex manifold. The moduli space is the semi-universal deformation space for such complex structures.

More precisely: A Riemann surface Σ is S equipped with an (almost) complex structure. The relationship with 1 depends on the Poincaré uniformization theorem, which states that each compact Riemann surface of genus $p > 1$ can be represented as a quotient of H as in 1. Conversely, each quotient H/Γ as in 1 obviously inherits a complex structure from H, since Γ operates by complex automorphisms on H.

The moduli space M_p is then constructed as a universal space for variations of complex structures. This means that if N is a complex space fibering over some base B with the generic (=regular) fiber being a Riemann surface of genus p, we then obtain a holomorphic map $h : B_0 \to M_p$ where $B_0 \subset B$ are the points with regular fibers. In this manner, M_p, as a moduli space of complex structures, acquires a complex structure itself that is determined by the requirement that all these h be holomorphic. Ideally, we would also like to have a holomorphic map $h_{fine} : N_0 \to M_{p,fine}$, N_0 being the space of regular fibers in N, mapping the fiber over $q \in B_0$ to the fiber over $h(q)$ in $M_{p,fine}$, but this is not always possible due to the difficulties with Riemann surfaces with nontrivial automorphisms. More precisely, $M_{p,fine}$ does not exist as such. A slight modification, however, leads to such a fine moduli space; namely, we only need to equip our Riemann surfaces additionally with some choice of a root of the canonical bundle in order to prevent nontrivial automorphisms. This is called a level structure. This gives a finite ramified cover of M_p. That cover is free of singularities and then yields a fine moduli space. (The Teichmüller space briefly described above is also a singularity-free cover of the moduli space, but, in contrast to the fine moduli space just introduced, it is an infinite cover and therefore not amenable to the constructions and techniques of algebraic geometry.) It is more subtle to understand what happens at the singular fibers. Here, we need a suitable compactification $\overline{M}_p$ of M_p through certain singular Riemann surfaces. This, however, is better understood through the subsequent approaches to M_p described below.

This construction is useful because, for example, it allows a geometric proof of the theorems of Arakelov-Parshin and Manin on the finiteness of the number of such fibrations of genus p over a given compact base B and the finiteness of the number of holomorphic sections of any given such fibration, see [69]. The idea is to show that because of the geometric properties of M_p and $\overline{M}_p$, there can only be finitely many such holomorphic maps $h : B \to \overline{M}_p$ or (after taking care of the above need to take finite covers) from N into a compactified fine moduli space.

3. A Riemann surface is an algebraic curve, described by homogeneous polynomial equations. The moduli space is the space of coefficients of these polynomials modulo projective automorphisms.

More precisely, a Riemann surface can be locally described as the common zero set of two homogeneous polynomials in three variables. The relationship with 2 depends on the Riemann–Roch theorem, which yields the existence of meromorphic functions.

Insertion: We briefly describe the relevant concepts. A line bundle L on Σ is given by an open covering $\{U_i\}_{i=1,...,m}$ of Σ and transition functions $g_{ij} \in \mathcal{O}^*(U_i \cap U_j)$ ($\mathcal{O}^*$ denoting the nonvanishing holomorphic functions) satisfying

$$g_{ij} \cdot g_{ji} \equiv 1 \quad \text{on } U_i \cap U_j \text{ for all } i, j, \tag{1.4.2}$$

$$g_{ij} \cdot g_{jk} \cdot g_{ki} \equiv 1 \quad \text{on } U_i \cap U_j \cap U_k \text{ for all } i, j, k. \tag{1.4.3}$$

Two line bundles L, L' with transition functions g_{ij} and g'_{ij}, resp., are called isomorphic if there exist functions $\phi_i \in \mathcal{O}^*(U_i)$ for each i with

$$g'_{ij} = \frac{\phi_i}{\phi_j} g_{ij} \quad \text{on each } U_i \cap U_j.$$

By multiplying transition functions we can define products of line bundles. The Abelian group of line bundles on Σ is called the Picard group of Σ, Pic(Σ). The Picard group Pic(Σ) is isomorphic to the group of divisors Div(Σ) modulo linear equivalence. (Divisors are finite formal sums $\sum n_\alpha p_\alpha$ with $n_\alpha \in \mathbb{Z}$, $p_\alpha \in \Sigma$. The addition in $\mathbb{Z}$ induces a group structure on these divisors. Divisors are linearly equivalent when their difference is the divisor defined by a meromorphic function. This is verified when one expresses a divisor D in terms of its local defining functions:

$$\left\{ (U_i, f_i) : \frac{f_i}{f_j} \in \mathcal{O}^*(U_i \cap U_j) \right\}.$$

The function f_i is meromorphic on U_i. When $p_\alpha \in U_i$, we require that f_i has a zero (pole) of order n_α at p_α if $n_\alpha > 0 (< 0)$. At all other points, f_i has to be holomorphic and nonzero.

We put

$$g_{ij} := \frac{f_i}{f_j} \quad \text{on } U_i \cap U_j$$

to define a line bundle, denoted by $[D]$.

Let L be a line bundle with transition functions g_{ij}. A holomorphic section h of L is given by a collection $\{h_i \in \mathcal{O}(U_i)\}$ of holomorphic functions on U_i satisfying

$$h_i = g_{ij} h_j \quad \text{on } U_i \cap U_j.$$

The zeros of a holomorphic section of a line bundle L define an effective (i.e., all $n_i > 0$) divisor E, and when $L = [D]$, that divisor is linearly equivalent to D, that is, $E - D$ is the divisor of a meromorphic function. In general, the zeros and poles of a meromorphic section of L define a divisor D with $[D] = L$. The degree of a divisor is the sum of its coefficients, and from this one then defines also the degree of the line bundle $[D]$. Thus, the degree of a line bundle counts the zeros minus the poles of a meromorphic section.

The Riemann–Roch theorem for line bundles is then

Theorem 1.3 *Let L be a line bundle on the compact Riemann surface Σ of genus p. Then the dimension of the space of holomorphic sections of L satisfies the relation*

$$h^0(L) = \deg L - p + 1 + h^0(K \otimes L^{-1}) \tag{1.4.4}$$

where K is the canonical bundle of Σ, that is, the line bundle of holomorphic 1-forms.

The equivalent formulation in terms of divisors replaces $h^0(L)$ by $h^0(D)$, the dimension of the space of effective divisors linearly equivalent to D.

Thus, the Riemann–Roch theorem can be viewed as an existence theorem for meromorphic functions, or, equivalently, for holomorphic sections of line bundles, whenever the right-hand side of (1.4.4) is positive. For example, $\deg K = 2p - 2$ and $h^0(K) = p$, $\deg K^2 = 2 \deg K = 4p - 4$ and $h^0(K^2) = 3p - 3$ for $p > 1$; K^2 is the line bundle whose sections are holomorphic quadratic differentials, that is, locally of the form $\varphi(z)dz^2$ with a holomorphic φ. Collections of holomorphic sections of a line bundle L define mappings into projective spaces because a change of the local representation of L multiplies them all by the same factor. One then needs sufficiently many independent sections to make such a map injective and thus to define an embedding of the Riemann surface into a projective space. In fact, one can show that every compact Riemann surface can be holomorphically embedded into $\mathbb{CP}^3$. Moreover, since by Chow's theorem every complex subvariety of $\mathbb{CP}^n$ is algebraic, our Riemann surface can then be represented by polynomial equations.

The relationship with 1 again goes via 2, that is, via the uniformization theorem. That theorem, however, is of a transcendental nature and thus outside the realm of algebraic geometry.

So, a Riemann surface becomes a (projective) algebraic variety in $\mathbb{CP}^3$, the zero set of algebraic equations. Such equations of a given degree can then be characterized by their coefficients. As automorphisms of $\mathbb{CP}^3$ lead to equivalent algebraic curves, one needs to divide these out. A difficulty emerges because the automorphism group of $\mathbb{CP}^3$ is not compact. Building upon the ideas of Hilbert, Mumford [83, 84] then developed geometric invariant theory to obtain the moduli space of algebraic curves. One then obtains the compactified Mumford–Deligne moduli space $\overline{M}_p$ as the moduli space of so-called stable curves, see [25]. As a moduli space of algebraic varieties, it is an algebraic variety itself, in agreement with the general principle.

4. A Riemann surface is a collection of branch points on the Riemann sphere S^2 with branching orders satisfying the Riemann–Hurwitz formula. The moduli space is obtained from those collections by factoring out automorphisms of S^2.

More precisely: Via some meromorphic function (whose existence again comes from the Riemann–Roch theorem), a Riemann surface is a branched cover of S^2, the Riemann sphere, which can also be identified with $\mathbb{CP}^1$. Again, we need to divide out automorphisms, this time those of S^2; they have the effect of moving the branch points around. This approach already led Riemann to count the number of moduli for Riemann surfaces of a given genus, that is, the dimension of the moduli space. This is explained in [51], for example.

5. A Riemann surface is a finite algebraic extension of the field of rational functions $\mathbb{C}(x)$ in one variable over $\mathbb{C}$.

From the algebraic representation in 3, one deduces that the field $k(\Sigma)$ of meromorphic functions on Σ is a finite algebraic extension of the field of rational functions $\mathbb{C}(x)$ in one variable over $\mathbb{C}$. More precisely,

$$k(\Sigma) \sim \mathbb{C}(x)[y]/P(x,y) \tag{1.4.5}$$

for some irreducible polynomial P. For example, an elliptic curve, that is, a Riemann surface of genus 1, can be described by a cubic polynomial

$$y^2 - x(x-1)(x-\lambda) \tag{1.4.6}$$

for some $\lambda \in \mathbb{C} - \{0, 1\}$. For $z \in \Sigma$, we let R_z be those meromorphic functions that are holomorphic at z. R_z is then a subring of $k(\Sigma)$ and has a unique maximal ideal given by those functions that vanish at z. This means that, conversely, we can start with the field $k(\Sigma)$ and define the points of Σ as the maximal ideals of local subrings of $k(\Sigma)$, and we may define a Riemann surface as a field of the form $\mathbb{C}(x)[y]/P(x, y)$ for some irreducible polynomial P. This encodes the functorial aspects: Let Σ_1, Σ_2 be compact Riemann surfaces, and let

$$\phi : k(\Sigma_2) \to k(\Sigma_1) \tag{1.4.7}$$

be a homomorphism whose restriction to $\mathbb{C}$ is the identity. Then there exists a unique holomorphic map

$$h : \Sigma_1 \to \Sigma_2 \tag{1.4.8}$$

with

$$\phi(f)(z) = f(h(z)) \tag{1.4.9}$$

for all $z \in \Sigma_1$ and all $f \in k(\Sigma_2)$.

This algebraic definition of a Riemann surface, which goes back to Dedekind and Weber, has the advantage that $\mathbb{C}$ can be replaced by any other algebraically closed field as the ground field. We may take finite fields $\mathbb{Z}_p$, and we can consider an algebraic equation $P(x, y)$ giving our Riemann surface as above, modulo p. Doing this for all prime numbers p simultaneously yields important insights into the algebraic properties of such equations, see [36], and this was at the heart of Faltings' proof of the Mordell conjecture [35]. We can also take, instead of $\mathbb{C}$, a field of meromorphic functions on some variety B, in order to obtain an algebraic curve over a function field. In more elementary terminology, we now consider a polynomial $P(x, y)$ whose coefficients depend on the variable $w \in B$. We thus obtain a family of Riemann surfaces as in 2, but now from an algebraic point of view. The unification of those two possibilities of considering varying ground fields (depending on a prime number p or on a variable w in some algebraic variety) leads to arithmetic algebraic geometry.

6. A Riemann surface Σ is (defined by) an Abelian variety with a principal polarization, its Jacobian, that can be identified as the group of divisors of degree 0 on Σ modulo linear equivalence or, equivalently, as the subgroup of the Picard group of line bundles of degree 0. Since not every principally polarized Abelian variety arises in this manner as the Jacobian of some Riemann surface, however, the moduli space of the latter is only a subvariety of the moduli space of principally polarized Abelian varieties. By considering periods of holomorphic 1-forms, we can associate to a Riemann surface a principally polarized Abelian variety, its Jacobian. By Torelli's theorem, each Riemann surface is determined by its Jacobian. This means that we can identify a Riemann surface with this Abelian variety, and the space of Riemann surfaces becomes a subspace of the moduli space of principally polarized Abelian varieties. What is not so nice about this is that the solution of the Schottky problem, that is, the question of characterizing those Abelian varieties that are Jacobians of Riemann surfaces, is rather complicated [97].

Insertion: We explain the above concepts in some more detail. $H^0(\Sigma, \Omega^{1,0})$ is the space of holomorphic 1-forms on our compact Riemann surface Σ, that is, the holomorphic sections of the canonical bundle K. Thus, $h^0(\Omega^1) := h^0(K) = \dim_\mathbb{C} H^0(\Sigma, \Omega^{1,0}) = p$ by Riemann–Roch.

Let $\alpha_1, \ldots, \alpha_p$ be a basis of $H^0(\Sigma, \Omega^{1,0})$, and $a_1, b_1, \ldots, a_p, b_p$ a canonical homology basis for Σ. Then the period matrix of Σ is defined as

$$\begin{pmatrix} \int_{a_1} \alpha_1 & \cdots & \int_{b_p} \alpha_1 \\ \vdots & & \vdots \\ \int_{a_1} \alpha_p & \cdots & \int_{b_p} \alpha_p \end{pmatrix}.$$

The column vectors of π,

$$P_i := \left(\int_{a_i} \alpha_1, \ldots, \int_{a_i} \alpha_p \right) \quad \text{and} \quad P_{i+p} := \left(\int_{b_i} \alpha_1, \ldots, \int_{b_i} \alpha_p \right), \quad i = 1, \ldots, p,$$

are called the periods of Σ. $P_1, \ldots, P_{2p}$ are linearly independent over $\mathbb{R}$ and thus generate a lattice

$$\Lambda := \{ n_1 P_1 + \cdots + n_{2p} P_{2p}, \ n_j \in \mathbb{Z} \}$$

in $\mathbb{C}^p$.

Definition 1.8 The Jacobian variety $J(\Sigma)$ of Σ is the torus $\mathbb{C}^p / \Lambda$.

For each z_0 in Σ, we have the Abel map (a holomorphic embedding)

$$j : \Sigma \to J(\Sigma)$$

with

$$j(z) := \left(\int_{z_0}^z \alpha_1, \ldots, \int_{z_0}^z \alpha_p \right) \mod \Lambda.$$

Here $j(z)$ is independent of the choice of the path from z_0 to z, since a different choice changes the vector of integrals only by an element of Λ.

By the theorems of Abel and Jacobi, we obtain an isomorphism φ from the group $\mathrm{Pic}^0(\Sigma)$ of line bundles of degree 0, that is, from the group $\mathrm{Div}^0(\Sigma)$ of divisors of degree 0 modulo linear equivalence, into the Jacobian $J(\Sigma)$ by writing a divisor D of degree 0 as

$$D = \sum_\nu (z_\nu - w_\nu),$$

where $z_\nu, w_\nu \in \Sigma$ are not necessarily distinct, and putting

$$\varphi(D) := \left(\sum_\nu \int_{w_\nu}^{z_\nu} \alpha_1, \ldots, \sum_\nu \int_{w_\nu}^{z_\nu} \alpha_p \right) \mod \Lambda.$$

7. A Riemann surface is a conformal structure on S, that is, a possibility to measure angles. Equivalently, it is an isometry class of Riemannian metrics modulo conformal factors. The moduli space is obtained by dividing the space of all Riemannian metrics on S by isometries and conformal changes. More

precisely: As already discovered by Gauss, a two-dimensional Riemannian manifold defines a conformal structure, that is, a Riemann surface. Different Riemannian
metrics can lead to the same conformal structure, and so we need to divide out
such equivalences. This is the approach of Tromba and Fischer, see [101]. Thus, we
consider the space R_p of all Riemannian metrics on S. As a space of Riemannian
metrics, it carries itself a Riemannian metric. If g is a Riemannian metric on S and
$h : S \rightarrow S$ is a diffeomorphism, h^*g is isometric to g via h. Thus, we need to divide
out the action of the diffeomorphism group D_p of S. It acts isometrically on R_p
equipped with its Riemannian metric. Moreover, when we multiply a given metric
g by some positive function λ, the metric λg leads to the same conformal structure
as g. Such multiplication by a positive function, however, does not induce an isometry of R_p (and this is at the heart of the anomalies in string theory that ultimately
force a particular dimension (26 in bosonic string theory)).

Insertion: Some details: Let $g \in R_p$ be some Riemannian metric on S. Suppress-
ing the issues of the precise regularity class of the objects encountered, the tangent
space $T_g R_p$ is given by symmetric 2×2 tensors $h = (h_{ij})$. Each such h can be
decomposed into its trace and trace-free parts:

$$h = \rho g + h' \qquad \rho : S \rightarrow \mathbb{R}, \tag{1.4.10}$$

$$h'_{ij} = h_{ij} - \frac{1}{2} g_{ij} g^{k\ell} h_{kl}. \tag{1.4.11}$$

The decomposition (1.4.10) is orthogonal w.r.t. the natural Riemannian structure on
$T_g R_p$:

$$((h_{ij}), (\ell_{ij}))_{g,\kappa} := \int (g^{ijkm} + \kappa g^{ij} g^{km}) h_{ij} \ell_{km} \sqrt{\det g}\, dz^1\, dz^2 \tag{1.4.12}$$

with $\kappa > 0$ and

$$g^{ijkm} := \frac{1}{2} (g^{ik} g^{jm} + g^{im} g^{jk} - g^{ij} g^{km}).$$

Since the value of κ will make no difference for us, we put $\kappa = \frac{1}{2}$ so that (1.4.12)
becomes

$$((h_{ij}), (\ell_{ij}))_g := \int_S g^{ij} g^{km} h_{ik} \ell_{jm} \sqrt{\det g}\, dz^1\, dz^2. \tag{1.4.13}$$

As it stands, this is only a weak Riemannian metric on the infinite-dimensional space
R_p, as (1.4.13) yields only an L^2-product, but Clarke [20] showed that it becomes
a metric space with respect to the distance function induced by the Riemannian
product of (1.4.13). (The completion of this metric space is identified in [19].)

From (1.4.13), we see that the Riemannian metric $(., .)_g$ on $T_g R_p$ is invariant
under the action of the diffeomorphism group, but not under conformal transformations.

In order to get rid of the ambiguity of the conformal factor, we need to find a suitable slice in R_p transversal to the conformal changes. By Poincaré's theorem, any
Riemannian metric on our surface S of genus $p > 1$ is conformally equivalent to

a unique hyperbolic metric, that is, S becomes a quotient H/Γ as above. This metric has constant curvature -1. With some differential geometry, one verifies that -1 is a regular value of the curvature functional, and so, by the implicit function theorem, the hyperbolic metrics yield a regular slice. Thus, we obtain the moduli space M_p as the space $R_{p,-1}$ of metrics of curvature -1 divided by the action of D_p. In this way, the geometric structures on R_p induce corresponding geometric structures on M_p as described in Tromba's book [101]. R_p is the space of symmetric, positive definite 2×2 tensors (g_{ij}) on S. As already explained, a tangent vector to R_p is then a symmetric 2×2 tensor (h_{ij}), not necessarily positive definite. It is orthogonal to the conformal multiplications when it is trace-free, and it is orthogonal to the action of D_p when it is divergence-free. Such a trace- and divergence-free symmetric tensor then can be identified with a holomorphic quadratic differential on the Riemann surface.

Insertion: Some details: We recall the decomposition

$$h = \rho g + h', \qquad (1.4.14)$$

where h' is trace-free. As we have seen, this decomposition is orthogonal w.r.t. the natural Riemannian metric on $T_g R_p$. In particular, since we only want to keep those directions that are orthogonal to conformal reparametrizations, we only need to consider the trace-free part h'. We next consider the infinitesimal action of the diffeomorphism group, with the aim of determining those h' that are orthogonal to the action of that group as well. For that purpose, let $(\varphi_t) \subset D_p$, $\varphi_0 = id$, be a smooth family of diffeomorphisms, generated by the vector field

$$V(z) := \frac{d}{dt}\varphi_t(z)_{|t=0}. \qquad (1.4.15)$$

The infinitesimal change of the metric g under (φ_t) is then given by

$$\frac{d}{dt}(\varphi_t^* g)_{|t=0}. \qquad (1.4.16)$$

(This is the Lie derivative $L_V g$ of the metric in the direction of the vector field V.) With ∇ denoting the covariant derivative for the metric g,

$$\frac{d}{dt}((\varphi_t^* g)_{|t=0})_{ij} = g_{ik}(\nabla_{\frac{\partial}{\partial z^j}} V)^k + g_{jk}(\nabla_{\frac{\partial}{\partial z^i}})^k$$

$$= g_{ij,k} V^k + g_{ik} V_{zj}^k + g_{jk} V_{zi}^k. \qquad (1.4.17)$$

In the above decomposition of R_p, the directions corresponding to conformal changes are given by the tensors ρg, whereas those representing D_p are of the form (1.4.17). It remains to identify the Teichmüller directions, i.e., those that are orthogonal to the preceding two types.

Our computations simplify considerably if we use conformal coordinates so that the metric (g_{ij}) is of the form

$$g_{ij}(z) = \lambda^2(z)\delta_{ij}. \qquad (1.4.18)$$

If a symmetric tensor h'' is orthogonal to all multiples ρg of g, it has to be trace-free. If it is orthogonal to all tensors that arise from the infinitesimal action of the diffeomorphism group, that is, of type (1.4.16), we get, using the symmetry of h''

$$0 = \int g^{ij} g^{kl} h''_{ik} (g_{j\ell,m} V^m + 2g_{jm} V^m_{z\ell}) \sqrt{\det g} \, dz^1 \, dz^2$$

$$= \int \frac{1}{\lambda^2} h''_{ik} \left(\delta_{ik} \left(\frac{\partial}{\partial z^m} \lambda^2 \right) V^m + 2\lambda^2 V^i_{zk} \right) dz^1 \, dz^2$$

$$= \int 2 h''_{ik} V^i_{zk} \, dz^1 \, dz^2, \quad \text{since } h'' \text{ is traceless.}$$

If this holds for all vector fields V, we conclude

$$\frac{\partial}{\partial z^k} h''_{ik} = 0 \quad \text{for } i = 1, 2. \tag{1.4.19}$$

This means that h''_{ik} is divergence free.

Thus, h'' is symmetric, trace-free, and divergence free. These conditions can be interpreted in a more concise manner as follows:

Being symmetric and trace-free, h'' is of the form

$$\begin{pmatrix} h''_{11} & h''_{12} \\ h''_{12} & h''_{22} \end{pmatrix} =: \begin{pmatrix} u & v \\ v & -u \end{pmatrix}.$$

Being divergence free, this tensor then has to satisfy

$$u_{z^1} = -v_{z^2}, \qquad u_{z^2} = v_{z^1}.$$

Thus, $u - iv$ is holomorphic, or, as a tensor,

$$h'' = u(dz^1)^2 - u(dz^2)^2 + 2v \, dz^1 \, dz^2$$

$$= \text{Re}((u - iv)(dz^1 + i dz^2)^2) \tag{1.4.20}$$

is the real part of a holomorphic quadratic differential

$$\phi \, dz^2 = (u - iv) \, dz^2.$$

Thus, we have identified the tangent directions of R_p that correspond to nontrivial deformations of the complex structure as the (real parts of) holomorphic quadratic differentials on the Riemann surface defined by (S, g).

Thus, the cotangent[11] space of M_p at a point representing a Riemann surface Σ is given by the holomorphic quadratic differentials on Σ. (This issue will be taken up again in Sect. 2.4 from a different point of view that also clarifies the relation between tangent and cotangent directions to the moduli space.) The complex dimension of this space is $3p - 3$ the Riemann–Roch theorem. M_p then also inherits a Riemannian structure from that of R_p. The induced metric is the Petersson–Weil metric originally introduced in the context of approach 1. In a more abstract

---

[11]It is not very transparent from our preceding considerations that we have constructed the cotangent and not the tangent space, but a careful accounting of the transformation behaviors can clarify this issue.

framework, the so-called L^2-geometry of moduli spaces is investigated in [67]. Let $\Phi\,dz^2 = (u_1 - iv_1)\,dz^2$ and $\Psi\,dz^2 = (u_2 - iv_2)\,dz^2$ be two such differentials. Let $\rho^2(z)\,dz\,d\bar{z}$ be the hyperbolic metric. Then their Petersson–Weil product is

$$(\Phi\,dz^2, \Psi\,dz^2)_g = 2\int (u_1 u_2 + v_1 v_2) \cdot \frac{1}{\rho^2(z)}\,dz\,d\bar{z} = 2\,\mathrm{Re}\int \Phi\bar{\Psi}\frac{1}{\rho^2(z)}\,dz\,d\bar{z}. \tag{1.4.21}$$

We have now listed seven rather different approaches for defining what a Riemann surface is. It is a very remarkable and profound fact that all these approaches give fully compatible structures on the moduli space M_p. In each of them, one can construct a complex structure on M_p, and they all agree, and together with the Petersson–Weil metric, one then finds a Kähler structure on M_p.

Nevertheless, some remarks are in order here:

- From an algebraic point of view, the hyperbolic metric is a transcendental object and should be replaced by an algebraic one. There are also certain other natural metrics on a Riemann surface, like the Bergmann metric obtained from an L^2-orthonormal basis of holomorphic 1-forms, that is, the metric induced by embedding the Riemann surface into its Jacobian, or the Arakelov metric defined from an asymptotic expansion of the Green function, a rather natural object in string theory. One may replace the hyperbolic metric in (1.4.21) by another metric uniquely associated to each Riemann surface and still obtain a natural Riemannian metric on M_p. First steps in the direction of a systematic investigation have been done in [54, 55]. For more recent results in this direction, see [58, 59]. Let us briefly describe some of these constructions here. The Bergmann metric is given by

$$\rho_B^2\,dz \wedge d\bar{z} := \sum_{i=1}^{p} \theta_i \wedge \bar{\theta}_i \tag{1.4.22}$$

where the θ_i are an L^2-orthonormal basis of the space of holomorphic 1-forms on Σ, that is,

$$\frac{i}{2}\int_{\Sigma} \theta_i \wedge \bar{\theta}_j = \delta_{ij}. \tag{1.4.23}$$

Equivalently, the metric is induced from the flat metric on the Jacobian $J(\Sigma)$ via the period map $j : \Sigma \to J(\Sigma)$. This latter description also shows that it does not depend on the choice of orthonormal basis—which, of course, is also readily checked directly. Moreover, the expression for the Bergmann metric is indeed positive definite, that is, it defines a metric, or equivalently, the derivative of the period map j has maximal rank. This follows from the fact that there is no point on Σ where all holomorphic 1-forms vanish simultaneously; this can be deduced from the Riemann–Roch theorem.

The Arakelov metric (references are [4, 18]) $\gamma^2 dz d\bar{z}$ is characterized by the property that its curvature is proportional to the Bergmann metric,

$$\frac{\partial^2}{\partial z \partial \bar{z}} \log \gamma = c_p \rho_B^2, \tag{1.4.24}$$

for some constant c_p that depends only on p and can, of course, be explicitly computed.[12] Alternatively, it is given in terms of an asymptotic expansion of the Green function of the Bergmann metric,

$$\log \gamma(z) = -\lim_{w \to z} (2\pi G(z, w) - \log|z - w|), \tag{1.4.25}$$

with G satisfying

$$\frac{\partial^2}{\partial z \partial \bar{z}} G(z, w) = \frac{i}{2} \delta_w(z) + c_p \rho_B^2, \tag{1.4.26}$$

where δ_w is the Dirac functional supported at w, plus the normalization[13]

$$\int_\Sigma G(z, w) \frac{i}{2} \rho_B^2 dz \wedge d\bar{z} = 0. \tag{1.4.27}$$

The Green function is regular for $z \neq w$ and becomes $-\infty$ at $z = w$. $\exp 2\pi G(z, w)$ vanishes to first-order at $z = w$. The first term in the expansion of $\exp 2\pi G(z, w)$ is the universal term $|z - w|$, while the next one, $\gamma(z)$, encodes the geometry of the Riemann surface Σ.

If Δ_B is the Laplace operator for the Bergmann metric, and if $\phi_0, \phi_1, \ldots$ is an L^2-orthonormal basis of eigenfunctions with eigenvalues $0 = \lambda_0 < \lambda_1 \leq \lambda_2 \ldots$, then the Green function is given by the expansion

$$G(z, w) = \sum_{j=1}^\infty \frac{1}{\lambda_j} \phi_j(z) \overline{\phi_j(w)}. \tag{1.4.28}$$

In fact, one can perform this construction of the Green function and the associated metric on the basis of any conformal metric on Σ in place of the Bergmann one. Arakelov discovered, however, that the Bergmann metric is distinguished here by the following property: When we use the Green function of a metric g to define a metric on the canonical bundle K by putting

$$\|dz\|(z_0) := \left(\lim_{z \to z_0} \frac{\exp 2\pi G(z, z_0)}{|z - z_0|} \right)^{-1}, \tag{1.4.29}$$

where the absolute value on the right-hand side is taken w.r.t. to local coordinates, that is, in $\mathbb{C}$, then the curvature of this metric on K is a multiple of g if and only

---

[12] In the sequel, c_p will denote a generic such constant whose value can change between formulas.

[13] The characterization of the Arakelov metric in terms of its curvature likewise needs an additional normalization to fully determine it.

if we started with the Bergmann metric. In other words, we have the formula

$$\frac{1}{2\pi i} \frac{\partial^2}{\partial z \partial \bar{z}} \log \|s\|^2 dz \wedge d\bar{z} = c_p \rho_B^2 dz \wedge d\bar{z} \qquad (1.4.30)$$

for any locally nonvanishing holomorphic section s of K, and this is no longer valid for other metrics g used to construct a Green function.

More generally, for a line bundle L over Σ with transition functions g_{ij}, a Hermitian metric λ^2 on L is a collection of positive, smooth, real-valued functions λ_i^2 on U_i with

$$\lambda_j^2 = \lambda_i^2 g_{ij} \overline{g_{ij}} \quad \text{on } U_i \cap U_j. \qquad (1.4.31)$$

The norm of a section h of L given by the local collection h_i is then defined via

$$\|h(z_0)\|^2 := \frac{|h_i(z_0)|^2}{\lambda_i^2(z_0)} \quad \text{for } z \in U_i. \qquad (1.4.32)$$

The curvature or first Chern form is given by

$$c_1(L, \lambda^2) := \frac{1}{2\pi i} \frac{\partial^2}{\partial z \partial \bar{z}} \log \|h\|^2 dz \wedge d\bar{z} \qquad (1.4.33)$$

for any meromorphic section h and local coordinates z, and this is independent of the choices of h and z. Arakelov called a Hermitian line bundle L admissible w.r.t. a metric $\rho^2 dz \wedge d\bar{z}$ on Σ if

$$c_1(L, \lambda^2) = \deg L \rho^2 \, dz \wedge d\bar{z}. \qquad (1.4.34)$$

Let $z_0 \in \Sigma$, and let z be local coordinates mapping z_0 to 0. We can put a Hermitian metric on the line bundle $[z_0]$ by defining the norm of the local section z in a neighborhood of z_0 as

$$|z|(z_1) = \exp G(z_1, z_0). \qquad (1.4.35)$$

This metric is then admissible for the Bergmann metric. So, what is special about the Bergmann metric here is that if we start the construction of the Arakelov metric from the Green function of that metric then the curvature formula recovers that metric. This only holds for the Bergmann metric and not for any other one.

- We noted in 6 that we can inject the moduli space M_p of Riemann surfaces of genus p into the moduli space A_p of principally polarized Abelian varieties of dimension p. The latter also carries a natural (locally Hermitian symmetric) metric. Since the map $j : M_p \to A_p$, while being injective by Torelli's theorem, is not of maximal rank everywhere, the pullback of that metric via j has some singularities. Also, its behavior is qualitatively different from that of the Weil–Petersson metric, as will become clear below when we investigate degenerations of Riemann surfaces and their associated Jacobians.

1.4.3 Compactifications of Moduli Spaces

Some of the preceding approaches also naturally lead to compactifications of M_p.

1. We already mentioned the Mumford–Deligne compactification $\overline{M}_p$ as an algebraic variety. It consists of so-called stable curves, that is, possibly singular curves, but with a finite automorphism group. The sphere with no, one, or two punctures and the torus are thereby excluded. This is necessary for the Hausdorff property.

   The difficulty here can be seen from the following easy example: We consider annuli, and by the uniformization theorem, each annulus is characterized by a single modulus, a real number $0 < r < 1$; that is, it is conformally equivalent to an annulus

$$A_r := \{z \in \mathbb{C} : r < |z| < 1\}. \tag{1.4.36}$$

Thus, the moduli space of annuli is $(0, 1)$. It seems obvious how to compactify it, namely by simply adding the boundary points $r = 0$ and $r = 1$. Now $r = 1$ does not correspond to a Riemann surface anymore, and so this is not a good limit. The annulus A_r, however, is conformally equivalent to the annulus

$$A_r' := \frac{1}{1-r}A_r = \left\{z \in \mathbb{C} : \frac{r}{1-r} < |z| < \frac{1}{1-r}\right\}, \tag{1.4.37}$$

which for $r \to 1$ converges to an infinite strip, that is, the limit can be identified with $\{x + iy \in \mathbb{C} : 0 < y < 1\}$. The boundary point $r = 0$ seems harmless because it simply corresponds to the punctured disk

$$D^* = \{z \in \mathbb{C} : 0 < |z| < 1\}. \tag{1.4.38}$$

However, the annulus A_r is also conformally equivalent to the annulus

$$A_r'' := \frac{1}{\sqrt{r}}A_r = \left\{z \in \mathbb{C} : \sqrt{r} < |z| < \frac{1}{\sqrt{r}}\right\}, \tag{1.4.39}$$

and if we now let r tend to 0, the limit is the punctured plane

$$\mathbb{C}^* = \{z \in \mathbb{C} : z \neq 0\}, \tag{1.4.40}$$

which is not conformally equivalent to the punctured disk D^*. Thus, from the same limit $r \to 0$, we obtain two different limits, D^* and $\mathbb{C}^*$, and therefore, we lose the Hausdorff property. Mumford's insight was that this problem essentially arises from the fact that the putative limit $\mathbb{C}^*$ has a noncompact automorphism group. In fact, its automorphism group contains all transformations of the form $z \to \lambda z$ for any $\lambda \in \mathbb{C}^*$. Mumford's theory then declared such limits as unstable and disallowed them. The problem of the noncompact automorphism group, however, will re-emerge later when we consider conformally invariant variational problems. The essential point is the following: We consider any Riemann surface Σ and choose local coordinates z in the open unit disk $U = \{z \in \mathbb{C} : |z| < 1\}$ around some point p_0, such that p_0 corresponds to 0. We then replace the coordinate z by $z_\lambda := \lambda z \in \lambda U = \{z \in \mathbb{C} : |z| < \lambda\}$. When we let $\lambda \in \mathbb{R}$ tend to

∞, we obtain $z_\infty \in \mathbb{C}$, but these are not coordinates for a local neighborhood of p_0 anymore because any fixed $z_\infty \in \mathbb{C}$ now corresponds to p_0 itself. In a sense to be made precise, they thus parametrize an infinitesimal neighborhood of p_0. We can compactify this infinitesimal coordinate patch $\mathbb{C}$ by adding the point at ∞ to obtain the sphere S^2. Thus, we have created a nontrivial Riemann surface, the sphere S^2, by blowing up a neighborhood of our point $p_0 \in \Sigma$. Again, if we allowed such processes in the construction of the moduli space, we would need to consider the union of Σ and S^2 as a limit of the constant sequence Σ. (As this so-called "bubbling off" can be repeated, we should then even allow for infinitely many blown-up spheres.) At this point, as mentioned, this can simply be excluded by fiat, but the situation changes when these blown-up spheres carry some additional data, for example some part of the Lagrangian action in a variational problem.

2. We recall from 2 in Sect. 1.4.2 that if N is a complex space fibering over some base B with the generic (=regular) fiber being a Riemann surface of genus p, then we obtain a holomorphic map $h : B_0 \to M_p$ where $B_0 \subset B$ are the points with regular fibers. The fibers over $B_1 := B \backslash B_0$ are then singular, and we hope to extend h across B_1, that is, obtain a holomorphic map $h : B \to \bar{M}_p$. Certain difficulties arise here from the possibility that not all such singular fibers in a holomorphic family need to be stable in the sense of Mumford. Thus, in particular, we cannot expect that the image of some point in B_1 is given by the complex structure of that singular fiber. Nevertheless, after lifting to finite covers so that the quotient singularities of M_p disappear, one can extend h to a holomorphic map $h : B \to \bar{M}_p$. This depends on certain hyperbolicity properties coming from the negative curvature of the Weil–Petersson metric on M_p that lead to general extension properties for holomorphic maps, see [69].

3. While the preceding is a global aspect, one also has a convenient local model for degenerations of Riemann surfaces within 2. We consider two unit disks $D_1 = \{z \in \mathbb{C} : |z| < 1\}$ and $D_2 = \{w \in \mathbb{C} : |w| < 1\}$. For $t \in \mathbb{C}$, $|t| < 1$, we remove the interior disks $\{|z| \leq |t|\}$, $\{|w| \leq |t|\}$ and glue the rest by identifying z with w by the equation $zw = t$ to obtain an annular region A_t. For $t \to 0$, A_t degenerates into the union of the two disks D_1, D_2 joined at the point $z = w = 0$. This is the local model for degeneration. The connection with the consideration of families as advocated in the preceding item of course comes from considering the smooth two-dimensional variety $N := \{(z, w, t) : zw - t = 0, |z|, |w|, |t| < 1\}$ for which (z, w) yield global coordinates. N fibers over the base $B := \{t : |t| < 1\}$, with a single singular fiber over $B_0 = \{0\}$.

This local model is easily implemented in the context of compact Riemann surfaces as follows. We let Σ_0 be either a connected Riemann surface of genus $p - 1 > 0$ with two distinguished points x_1, x_2, called punctures, or the disjoint union of two Riemann surfaces Σ^1, Σ^2 of genera $p_1, p_2 > 0$ with $p_1 + p_2 = p$ and one puncture $x_i \in \Sigma^i$ each. We choose disjoint neighborhoods U_1, U_2 of the punctures and local coordinates $z : U_1 \to D_1$, $w : U_2 \to D_2$ with $z(x_1) = 0$, $w(x_2) = 0$. By performing the above grafting process on the coordinate disks D_1 and D_2, we obtain a Riemann surface Σ_t of genus p for $t \neq 0$. The correspondence

$$t \mapsto \Sigma_t \tag{1.4.41}$$

induces a map of $D^* = \{t \in \mathbb{C} : 0 < |t| < 1\}$ onto a complex curve in the moduli space M_p which extends to a map from $D = \{t \in \mathbb{C} : |t| < 1\} = D^* \cup \{0\}$ into the compactification $\overline{M}_p$. (Because of the genus restrictions imposed, these degenerations all yield stable curves.)

4. Approaches 1 and 7 suggested looking at the moduli space of hyperbolic metrics. A hyperbolic metric on a compact surface can degenerate into a noncompact but complete hyperbolic metric of finite area with cusps. In the local model described in the previous item, this looks as follows. On the annulus $A_t = \{|t| < z < 1\}$, we have the hyperbolic metric

$$\frac{dz \wedge d\bar{z}}{|z|^2 \log^2 |z|} \left(\frac{\pi \frac{\log|z|}{\log|t|}}{\sin(\pi \frac{\log|z|}{\log|t|})} \right)^2 . \tag{1.4.42}$$

For $|t| \to 0$, this converges to the hyperbolic metric on the punctured disk $\{z : 0 < |z| < 1\}$ given by

$$\frac{dz \wedge d\bar{z}}{|z|^2 \log^2 |z|} . \tag{1.4.43}$$

This metric is complete at 0, that is, the cusp 0 is at infinite distance from the points in the punctured disk. Also, the area of every punctured subdisk $\{z : 0 < |z| < \rho\}, 0 < \rho < 1$ is finite.

For the hyperbolic metric (1.4.42) on the annulus A_t, the middle curve $|z| = \sqrt{|t|}$ is the shortest of all the concentric circles, hence a closed geodesic, denoted by c. The reflection $z \mapsto \frac{t}{z}$ is then an isometry leaving c fixed. Its length l goes to 0 as $t \to 0$, while its distance from the boundary $|z| = 1$ goes to ∞. Thus, as t goes to 0, the geodesic c degenerates into a point curve at infinite distance from the interior. Therefore, in geometric terms, the degeneration is described by pinching a closed geodesic on some annulus inside our Riemann surface equipped with the hyperbolic metric. In fact, a hyperbolic metric on an annulus that is symmetric about a closed geodesic is uniquely determined by the length of that geodesic. That means that the hyperbolic on the annulus A_t is induced by the metric of the Riemann surface Σ_t as described above. Thus, even though we have presented it here as a local model, it captures the essential global aspects.

This consideration of varying hyperbolic metrics leads to the same compactification $\overline{M}_p$ of M_p as a topological space, see [12]. The noncompact surfaces can be compactified as Riemann surfaces by adding a point at each cusp. We thus see that elements in the compactifying boundary of $\overline{M}_p$ correspond to surfaces of lower topological type with additional distinguished points, so-called punctures.

Insertion: The degeneration can also be described in terms of the generators of the discrete group Γ considered in 1. Since hyperbolic elements are characterized by $|a+d| > 2$ and parabolic ones by $|a+d| = 2$, the relevant degeneration is one where we have a sequence Γ_n of surface groups with hyperbolic elements γ_n converging to a parabolic element γ_0. An example is the sequence of hyperbolic transformations

$$\gamma_n : z \mapsto \frac{(1 + \frac{1}{n})z + \frac{1}{n}}{z + 1} \tag{1.4.44}$$

converging to the parabolic transformation

$$\gamma_0 : z \mapsto \frac{z}{z + 1}. \tag{1.4.45}$$

In the limit, the two fixed points of γ_n merge into the single fixed point 0 of γ_0. Also, the length of the invariant geodesic for γ_n approaches 0 as $n \to \infty$. Thus, again, the degeneration is described by pinching a closed geodesic on our Riemann surface equipped with the hyperbolic metric induced from H.

We now want to relate the geometric description of degeneration just established to the analytic model described previously. We first describe how to get from the analytic model to the geometric one. The behavior of the hyperbolic closed geodesic $|z| = \sqrt{|t|}$ for the hyperbolic metric (1.4.42) on the annulus A_t translates into the following picture for hyperbolic isometries of H. We consider the hyperbolic isometry $\gamma_\lambda : z \mapsto \lambda z$ for some $\lambda > 1$. This leaves the imaginary axis in H invariant, and so its image on the quotient H/Γ by the group Γ generated by γ_λ is a closed geodesic of length $\int_1^\lambda \frac{dy}{y} = \log \lambda$. Via $z \mapsto \log \lambda \exp(\frac{2\pi i}{\log \lambda}(\log(-iz) + \log \lambda))$, H/Γ is mapped onto $\mathbb{C}^*$, and the closed geodesic is mapped onto the circle $|w| = \log \lambda$.

In order to see how the geometric model can be translated into the analytic one, one uses the collar lemma, which says that if $\Sigma = H/\Gamma$ is a compact Riemann surface with a simple[14] closed geodesic c of length l, then Σ contains an annular region, called a collar, about c isometric to A_t with the hyperbolic metric A_t, c corresponding to the middle curve $|z| = \sqrt{|t|}$. The boundary curves of the collar then are at a distance from c of at least $\operatorname{arcsinh}(\frac{1}{\sinh(l/2)})$ which goes to ∞ as $l \to 0$. Thus, we are in the local situation described by the analytic model.

In fact, a theorem of Mumford says that pinching a simple closed geodesic is the only way a sequence of compact Riemann surfaces $\Sigma_n = H/\Gamma_n$ of fixed genus p can degenerate. Namely, if the lengths of (simple) closed geodesics on Σ_n are uniformly bounded below, then after selection of a subsequence, Γ_n converges to a subgroup Γ_0 of $PSL(2, \mathbb{R})$ for which $\Sigma_0 = H/\Gamma_0$ is a compact Riemann surface of the same genus p.

5. Since we have equipped M_p in approach 7 with a Riemannian metric, the Petersson–Weil metric, we can study its compactification as a metric space. Again, as follows from the computations and estimates of Masur [80], this leads to the same $\overline{M}_p$ viewed as a topological space, see [107]. In particular, M_p is not a complete metric space, that is, the boundary $\overline{M}_p \backslash M_p$ is at finite distance from the interior. Moreover, when we approach that boundary orthogonally

---

[14]That is, non-self-intersecting.

along some curve c, the tangent directions orthogonal to c converge to boundary tangent directions. For a survey of some recent refinements of these results, see [108]. For the relation with the completion of the space R_p of Riemannian metrics, see [19].

6. As explained in 6 of Sect. 1.4.1, by Torelli's theorem, the correspondence between a Riemann surface and its Jacobian leads to an injective mapping from M_p into the moduli space A_p of principally polarized Abelian varieties of dimension p. A_p is a quotient H_p/Λ_p of the Siegel upper half space by a discrete group (H_1 is simply the Poincaré upper half plane, and H_p/Λ_p is then a higher-dimensional generalization of the modular curve $H/\mathrm{SL}(2, \mathbb{Z})$. H_p is the space of symmetric complex ($p \times p$) matrices with positive definite imaginary part. The discrete group Λ_p is $\mathrm{Sp}(2p, \mathbb{Z})$, the group of real ($2p \times 2p$) matrices M with integer entries that satisfy $MJM^t = J$ for $J = \begin{pmatrix} 0 & Id \\ -Id & 0 \end{pmatrix}$). It admits a compactification first studied by Satake. Baily [8] then studied the induced compactification $\overline{\overline{M}}_p$. This is different from $\overline{M}_p$ and, in fact, highly singular. It can be obtained from $\overline{M}_p$ by forgetting the positions of the punctures or cusps of the limiting Riemann surfaces in $\overline{M}_p$. This is useful for the study of minimal surfaces of varying topological type, see [60, 61, 68], because the punctures would correspond to removable singularities. We shall discuss this issue briefly below in our study of the Dirichlet integral, our fundamental action functional, see Sect. 2.4.

Also, in string theory, one ultimately wishes to extend the partition function over all possible genera, and one therefore needs some kind of universal moduli space that includes surfaces of all possible genera. The problem with the Mumford–Deligne compactification is that as the genus increases one gets surfaces with more and more punctures in the low boundary strata, in fact infinitely many in the limit of the genus going to infinity. This is avoided in the Satake–Baily compactification just described.

There is another issue of interest here: We have described the degeneration of a family of Riemann surfaces by pinching a closed geodesic, that is, letting its length shrink to 0. These geodesics can be topologically of two different kinds. The first possibility is that it corresponds to a nontrivial homology class. When we pinch such a geodesic to a point and compactify the resulting surface by inserting two points, in the limit we still have a connected surface, but of lower genus, and therefore its space of holomorphic 1-forms has a smaller dimension. Therefore, the limiting surface also has a Jacobian of smaller dimension, and so we move into the boundary of A_p. The other possibility is that we pinch a curve that is homologically trivial, i.e., a commutator in the fundamental group $\pi_1(\Sigma)$. If we pinch such a curve and again compactify by inserting two points, the resulting surface is disconnected, but the genus p is not lowered, that is, the sum $p_1 + p_2$ of the genera of the pieces Σ_1 and Σ_2 equals p. Therefore, also the dimension of the Jacobian is not lowered, and although we move to the boundary of the moduli space M_p, we stay inside the moduli space A_p. The Jacobian of our disconnected surface is simply the product of the Jacobians of the pieces Σ_1 and Σ_2. Of course, in order to substantiate these contemplations, we need to clarify in which sense the Jacobians of the family of degenerating surfaces converge to the Jacobian of the compactified limiting surface.

1.5 Supermanifolds

1.5.1 The Functorial Approach

We present here the abstract mathematical setting of supermanifolds. We consider a super vector space (over a ground field of characteristic 0, like $\mathbb{R}$ or $\mathbb{C}$)

$$W = W_0 \oplus W_1$$

that is $\mathbb{Z}/2\mathbb{Z}$ graded. Elements w of W_0 are called even, with parity $p(w) = 0$, those of W_1 odd, with parity $p(w) = 1$. Morphisms between super vector spaces are required to preserve the grading.

A super algebra A is a super vector space together with a product $A \otimes A \to A$ which is a morphism in the above sense. It is also required to be associative and to have a unit, in the ordinary sense.

Now, the important point about super objects is that whenever an operation changes the order of two odd elements, a minus sign is introduced. In this sense, the super algebra A is (super)commutative if for any two $a, b \in A$,

$$ab = (-1)^{p(a)p(b)}ba. \tag{1.5.1}$$

(Here and in the sequel, whenever the parity of an element enters a formula, that element is implicitly assumed to be of pure type, that is, either odd or even, but not a nontrivial sum of an odd and an even term. Generally, definitions are extended to inhomogeneous elements by linearity.)

The basic example of a commutative super algebra is a Grassmann algebra with generators $v_1, \ldots, v_N$ satisfying

$$v_i v_j = -v_j v_i \quad \text{for all } i, j \tag{1.5.2}$$

and thus, in particular,

$$v_i^2 = 0 \quad \text{for all } i. \tag{1.5.3}$$

Hence, every element of this Grassmann algebra can be expanded as

$$v = a_0 + \sum_{i=1}^{N} a_i v_i + \cdots + a_{12\ldots N} v_1 v_2 \cdots v_N. \tag{1.5.4}$$

Since the square of any generator vanishes by (1.5.4), the expansion terminates.

Similarly to (1.5.1), the rules defining Lie algebras pick up signs in the super context: The bracket of a super Lie algebra has to satisfy

$$[v, w] + (-1)^{p(v)p(w)}[w, v] = 0 \tag{1.5.5}$$

and the super Jacobi identity reads

$$[v, [w, u]] + (-1)^{p(v)(p(w)+p(u))}[w, [u, v]] + (-1)^{p(u)(p(v)+p(w))}[u, [v, w]] = 0. \tag{1.5.6}$$

We now consider a complex super vector space W. A real structure on W is given by a $\mathbb{C}$-antilinear automorphism

$$\kappa : W \to W$$

with

$$\kappa^2 w = (-1)^{p(w)} w. \tag{1.5.7}$$

This should be considered as complex conjugation. We point out, however, that on the odd part, we obtain a minus sign in (1.5.7). As an example, let us assume that over $\mathbb{R}$, the odd part W_1 has two generators ϑ_1 and ϑ_2; we may then put

$$\kappa(\vartheta_1) = \vartheta_2, \qquad \kappa(\vartheta_2) = -\vartheta_1. \tag{1.5.8}$$

A supersymmetric bilinear form[15] $(\cdot, \cdot)$ on W is given by a symmetric form $(\cdot, \cdot)_0$ on W_0 and an alternating form $(\cdot, \cdot)_1$ on W_1 with

$$(\kappa v, \kappa w)_i = \overline{(v, w)_i} \quad \text{for } i = 0 \text{ and } 1. \tag{1.5.9}$$

This implies

$$\overline{(v, \kappa v)_i} = (\kappa v, \kappa^2 v)_i = (-1)^i (\kappa v, v)_i = (v, \kappa v)_i, \tag{1.5.10}$$

that is

$$(v, \kappa v) \quad \text{is real for all } v \in W. \tag{1.5.11}$$

We may thus call the form $(\cdot, \cdot)$ positive if

$$(v, \kappa v) > 0 \quad \text{for all } v \neq 0. \tag{1.5.12}$$

We can then define $(v, \kappa v)$ as a "norm" on a complex super vector space. The point is that $(v, v) = 0$ if v is odd. We therefore need κ which is only meaningful if W is defined over $\mathbb{C}$ so that each odd coordinate has two real components, as in our example. In that example, we could put

$$(\vartheta_1, \vartheta_2) = 1. \tag{1.5.13}$$

Then ϑ_1 and ϑ_2 would both have "norm" 1.

If we have a complex super algebra A, we could then require that $\kappa(ab) = \kappa(a) \kappa(b)$. If we wish to also include non-commutative algebras, like matrix algebras with their complex conjugation, it seems preferable to take as the basis object a star-operation, a $\mathbb{C}$-antilinear isomorphism from A to the opposite algebra[16] satisfying

$$(ab)^* = (-1)^{p(a)p(b)} b^* a^*. \tag{1.5.14}$$

---

[15]Forms always take their values in $\mathbb{C}$.

[16]If the product in A of a and b is ab, the product in the opposite algebra is defined as $(-1)^{p(a)p(b)} ba$.

A Hermitian form $\langle \cdot, \cdot \rangle$ on a complex super vector space is $\mathbb{C}$-antilinear in the first variable, $\mathbb{C}$-linear in the second one and satisfies

$$\overline{\langle v, w \rangle} = (-1)^{p(v)p(w)} \langle w, v \rangle. \qquad (1.5.15)$$

We have, since $\langle \cdot, \cdot \rangle$ is assumed to be even, that

$$\langle v, w \rangle = 0 \quad \text{if } p(v) \neq p(w), \qquad (1.5.16)$$

and also

$$\langle v, v \rangle \in \mathbb{R} \quad \text{for } v \text{ even}, \qquad \langle v, v \rangle \in i\mathbb{R} \quad \text{for } v \text{ odd}. \qquad (1.5.17)$$

Then a super Hilbert space H is a super vector space with a Hermitian form satisfying

$$\langle v, v \rangle > 0 \quad \text{for } v \text{ even}, \qquad (1.5.18)$$

$$i^{-1} \langle v, v \rangle > 0 \quad \text{for } v \text{ odd} \qquad (1.5.19)$$

and for which the ordinary Hilbert space structure defined by

$$\begin{aligned} \langle\!\langle v, w \rangle\!\rangle &= \langle v, w \rangle & \text{for } v, w \text{ even}, \\ \langle\!\langle v, w \rangle\!\rangle &= i^{-1} \langle v, w \rangle & \text{for } v, w \text{ odd}, \\ \langle\!\langle v, w \rangle\!\rangle &= 0 & \text{for } v, w \text{ of different parities} \end{aligned}$$

is complete. In the present treatise, we shall be concerned only with finite-dimensional super Hilbert spaces,[17] and the completeness is not an issue then because finite-dimensional Euclidean spaces are always complete.

1.5.2 Supermanifolds

As for ordinary manifolds, there are several approaches to the definition of supermanifolds, and it is instructive to understand the relations between them. The standard model is $\mathbb{R}^{m|n}$ with even coordinates $(x^1, \ldots, x^m)$ and odd coordinates $(\vartheta^1, \ldots, \vartheta^n)$. Its sheaf of functions is $C^\infty[\vartheta^1, \ldots, \vartheta^n]$, the sheaf of commutative super algebras freely generated by odd quantities $\vartheta^1, \ldots, \vartheta^n$ over the sheaf C^∞ of smooth functions on $\mathbb{R}^m$. Since the square of any ϑ^j vanishes, they generate a nilpotent ideal in this sheaf.

The functions in $C^\infty[\vartheta^1, \ldots, \vartheta^n]$ then admit expansions in the nilpotent variables. To explain this, we first consider $x = (x^1, \ldots, x^m) \in U$ (open in $\mathbb{R}^m$) and $\xi = (\xi^1, \ldots, \xi^m)$ where the ξ^i are even nilpotent elements, i.e., of the form $\sum_{\alpha_1, \alpha_2} a_{\alpha_1, \alpha_2} \vartheta^{\alpha_1} \vartheta^{\alpha_2} +$ higher even-order terms, that is, the expansion starts with products of two ϑ^is. For a function that depends only on the even variables, we then require

$$F(x^1 + \xi^1, \ldots, x^m + \xi^m)$$

$$= \sum_\gamma \frac{1}{\gamma_1! \cdots \gamma_m!} \partial_{x^1}^{\gamma_1} \cdots \partial_{x^m}^{\gamma_m} F(x^1, \ldots, x^m)(\xi^1)^{\gamma_1} \cdots (\xi^m)^{\gamma_m} \qquad (1.5.20)$$

---

[17]Perhaps, one should better speak of super Euclidean spaces in that case.

where F as a function of $x = (x^1, \ldots, x^m)$ is of class $C^\infty(U)$. Alternatively, we can view this as the rule for extending or pulling back a function of the ordinary coordinates $x = (x^1, \ldots, x^m)$ to one of the coordinates $x + \xi = (x^1 + \xi^1, \ldots, x^m + \xi^m)$. When the function is also allowed to depend on the odd variables, we have the expansion

$$F(x^1 + \xi^1, \ldots, x^m + \xi^m, \vartheta^1, \ldots, \vartheta^n)$$

$$= \sum_\alpha \sum_\gamma \frac{1}{\gamma_1! \cdots \gamma_m!} \partial_{x^1}^{\gamma_1} \cdots \partial_{x^m}^{\gamma_m} F^\alpha(x^1, \ldots, x^m)(\xi^1)^{\gamma_1} \cdots (\xi^m)^{\gamma_m} \vartheta^{\alpha_1} \ldots \vartheta^{\alpha_k}$$

$$(1.5.21)$$

where the functions F^α are of class $C^\infty(U)$. In these expansions, we may also allow for functions F^α taking their values in a supercommutative algebra with unit in place of $\mathbb{R}$. Usually, these functions will then be even. We note that the expansions (1.5.20) and (1.5.21) contain a number of derivatives that depend on n. Since we want to reserve the flexibility to keep n variable, we must work with C^∞- instead of C^k-functions for some finite k.

There also exists a notion of (formal) integration, the Berezin integral, that inverts differentiation.

If F is only a function of one odd variable ϑ, we have

$$F(\vartheta) = a + b\vartheta \tag{1.5.22}$$

where a, b are constants, i.e., independent of ϑ. The integral of F w.r.t. ϑ is then defined by linearity and the basic rules

$$\int d\vartheta = 0, \qquad \int \vartheta \, d\vartheta = 1. \tag{1.5.23}$$

This makes the integral translation invariant, i.e. for an odd ε,

$$\int F(\vartheta + \varepsilon) d\vartheta = \int (a + b\vartheta + b\varepsilon) d\vartheta = b \int \vartheta \, d\vartheta = \int F(\vartheta) d\vartheta. \tag{1.5.24}$$

Similarly, for a function F of n odd variables $\vartheta^1, \ldots, \vartheta^n$,

$$F(\vartheta^1, \ldots, \vartheta^n) = \sum_\alpha b_\alpha \vartheta^\alpha \quad \begin{pmatrix} \text{with } \alpha = 0 \text{ or } \alpha = (\alpha_1, \ldots, \alpha_k) \\ 1 \le \alpha_1 < \alpha_2 \ldots < \alpha_k \le n \end{pmatrix} \tag{1.5.25}$$

the integral is computed via the rules

$$\int d\vartheta^i = 0, \qquad \int \vartheta^i d\vartheta^j = \delta_{ij}. \tag{1.5.26}$$

Thus, we have

$$\int b_\alpha \vartheta^\alpha d\vartheta^{\alpha_k} \cdots d\vartheta^{\alpha_1} = b_\alpha \quad \text{for } \alpha = (\alpha_1, \ldots, \alpha_k). \tag{1.5.27}$$

A supermanifold of dimension $m|n$ can be defined by an atlas whose local charts are open domains of $\mathbb{R}^{m|n}$, that is, subsets with sheaf of functions $C^\infty(U_0)[\vartheta^1, \ldots, \vartheta^n]$, where U_0 is an open subset of $\mathbb{R}^m$. In terms of functions, we are restricting the sheaf $\mathcal{C}^\infty[\vartheta^1, \ldots, \vartheta^n]$ to U_0. This will of course be the general procedure for defining sub-supermanifolds. Note that we are restricting the even coordinates $x^1, \ldots, x^m$ here, but not the odd ones. So, we have coordinate charts; coordinate transformations are then given by isomorphisms $f : U \to V$, U, V open in $\mathbb{R}^{m|n}$. Such an isomorphism is given by even functions $f^1, \ldots, f^m$ and odd functions $\phi^1, \ldots, \phi^n$. To be an isomorphism, f must be invertible, and the functions must be smooth, as always. And a morphism is invertible iff the underlying morphism defined by the $f^1, \ldots, f^m$ is invertible; the odd functions do not play a role for invertibility.

Based on this, if F is a function on the chart U, and if $f^1, \ldots f^m, \phi^1, \ldots, \phi^n$ are coordinate functions on our supermanifold, we can compute the values $F(f^1, \ldots, f^m, \phi^1, \ldots, \phi^n)$. We can therefore equivalently define a supermanifold as a topological space M_0 with a sheaf $\mathcal{O}_M$ of super ($\mathbb{R}$)-algebras that is locally isomorphic to $\mathbb{R}^{m|n}$. Functions on M are then sections of the structure sheaf $\mathcal{O}_M$. Morphisms between supermanifolds $f : M \to N$ are then morphisms of ringed spaces, that is continuous maps $f_0 : M_0 \to N_0$ with a morphism of sheaves of super algebras from $f_0^* \mathcal{O}_N$ to $\mathcal{O}_M$. The odd functions generate a nilpotent ideal J of $\mathcal{O}_M$, because the square of any odd coordinate is 0. The space M_0 with the sheaf $\mathcal{O}_M/J$ is then a smooth manifold of dimension m, called the reduced manifold M_r. A function f on M projects to a function f_r on M_r, that is, a smooth function on M_0. The sheaf morphism determines the function. In particular, the evaluation of an odd function at a point of M_0 always yields 0. This also means that any map from an ordinary manifold, that is, a supermanifold of dimension $m|0$, into one of dimension $0|n$ vanishes identically. This can be remedied through the functor of points approach to supermanifolds. For a supermanifold S, an S-point of a supermanifold M is a morphism $S \to M$. This construction is functorial in the sense that a morphism $\psi : T \to S$ induces a map from $M(S)$, the set of S-points of M, to $M(T)$ via $m \mapsto m \circ \psi$. Similarly, a morphism $f : M \to N$ induces $f_S : M(S) \to N(S)$, again functorially in S. In order to understand this more abstractly, we consider the so-called superpoints, the supermanifolds $\mathbb{R}^{0|n}$ defined as the space with structure sheaf $\mathbb{R}[\vartheta^1, \ldots, \vartheta^n]$ (with anticommuting ϑ^j, as always). Expressed differently, these are the supermanifolds $(\{\star\}, \Lambda_n)$ where $\{\star\}$ is an ordinary point and Λ_n is a Grassmann algebra of n generators. Then, see [93], these superpoints generate the category of finite-dimensional supermanifolds, that is, any such supermanifold is completely described by its superpoints. For supermanifolds M, N, one then defines (or, more precisely, shows the existence of) the inner Hom object $\underline{\text{Hom}}(M, N)$ satisfying

$$\text{Hom}(\mathbb{R}^{0|n}, \underline{\text{Hom}}(M, N)) = \text{Hom}(\mathbb{R}^{0|n} \times M, N) \qquad (1.5.28)$$

for all $n \in \mathbb{N}$ and then also

$$\text{Hom}(S, \underline{\text{Hom}}(M, N)) = \text{Hom}(S \times M, N) \qquad (1.5.29)$$

for all supermanifolds S. In this way, the space of morphisms $M \to N$ also becomes a functor: For a supermanifold S, a morphism $M \times S \to N$, that is, a morphism $M \to N$ depending on a parameter in S, is then an S-point of $\underline{\text{Hom}}(M, N)$. The morphisms $\mathbb{R}^{0|n} \times M \to N$ are then the superpoints of the supermanifold of morphisms $\underline{\mathrm{Hom}}(M, N)$ (in contrast to $\mathrm{Hom}(M, N)$ which is not a supermanifold, but rather the reduced space (see below) underlying the supermanifold $\underline{\mathrm{Hom}}(M, N)$).

In that way, we see that there also exist nontrivial odd functions on an ordinary manifold, say $\mathbb{R}$, even though their values vanish on all points of $\mathbb{R}$. To be concrete, consider $S = \mathbb{R}^{0|n}$. $\mathbb{R} \times S$ then has the sheaf $C^{\infty}(\mathbb{R}) \otimes \mathbb{R}[\vartheta^1, \ldots, \vartheta^n]$. Now take another space T that is odd like S, with sheaf $\mathbb{R}[\eta^1, \ldots, \eta^m]$. We consider a map $\psi : \mathbb{R} \times S \to T$, that is $\psi : \mathbb{R}^{1|n} \to \mathbb{R}^{0|m}$, given by

$$\mathcal{C}^{\infty}(T) = \mathbb{R}[\eta^1, \ldots, \eta^m] \to C^{\infty}(\mathbb{R}) \otimes \mathbb{R}[\vartheta^1, \ldots, \vartheta^n],$$

$$\eta^j \mapsto a_k^j(t)\vartheta^k$$

which we can also write as

$$\eta^j(t) = a_k^j(t)\vartheta^k.$$

Of course, this vanishes at all points of $\mathbb{R} = (\mathbb{R} \times S)_0$, but nevertheless it is a nontrivial morphism.

In the converse direction, let us take $n = 1$, i.e., consider $S = \mathbb{R}^{0|1}$, the space with sheaf $\mathbb{R}[\vartheta]$ ($\vartheta^2 = 0$), and a morphism

$$S \to M_0$$

into some ordinary manifold M_0. This is given by an algebra homomorphism

$$C^{\infty}(M_0) \to \mathbb{R}[\vartheta],$$

$$f \mapsto a_0(f) + a_1(f)\vartheta.$$

The homomorphism condition implies first that

$$a_0 : C^{\infty}(M_0) \to \mathbb{R},$$

$$f \mapsto a_0(f)$$

is an algebra homomorphism, and, as is easily derived, it is therefore given by the evaluation at some point $x \in M_0$, that is $a_0(f) = f(x)$. Secondly we obtain, using the homomorphism condition,

$$a_0(fg) + a_1(fg)\vartheta = (a_0(f) + a_1(f)\vartheta)(a_0(g) + a_1(g)\vartheta)$$

$$= f(x)g(x) + (f(x)a_1(g) + g(x)a_1(f))\vartheta,$$

which means that a_1 is a derivation over functions, that is, the derivative in the direction of some tangent vector $v_x \in T_x M_0$,

$$a_1(f) = v_x f.$$

Thus, we could view the super point S with its sheaf $\mathbb{R}[\vartheta]$ as an abstract (odd) tangent vector. The maps $S \to M_0$ correspond to points in the tangent bundle $T M_0$. If M is a general supermanifold, the same applies, except that we get a sign from the odd functions, that is,

$$a_1(fg) = a_1(f)g(x) + (-1)^{p(f)}f(x)a_1(g).$$

Thus, a_1 is an odd homomorphism from the local ring at x to $\mathbb{R}$.

In any case, we have a projection $M \to M_r$ from a supermanifold to its reduced manifold M_r. Conversely, by Batchelor's theorem, any smooth supermanifold is (non-canonically) isomorphic to one of the form $(M_r, \wedge^* V)$. Thus, we can obtain M from the smooth ordinary manifold M_r and a locally free module V over the sheaf $C^\infty(M_r)$; namely, M can be obtained as M_r with the sheaf $\wedge^* V$ graded by the exterior degree mod 2, and the inclusion of $C^\infty(M_r)$ into $\wedge^* V$ defines a morphism $M \to M_r$ that retracts the embedding of M_r into M. It is important to realize that these constructions are not canonical, since they are not invariant under automorphisms of M if $m \geq 1, n \geq 2$. Namely, simply consider $\mathbb{R}^{1|2}$ with coordinates $(x, \vartheta^1, \vartheta^2)$ and the automorphism

$$(x, \vartheta^1, \vartheta^2) \mapsto (x + \vartheta^1 \vartheta^2, \vartheta^1, \vartheta^2).$$

This example also shows us that decompositions of functions according to their degree are not invariant under automorphisms, and thus not invariant under coordinate transformations. Namely, if we have a function f that, in the coordinates $(x, \vartheta^1, \vartheta^2)$, only depends on x, and if we denote its expression in the new coordinates $(x + \vartheta^1 \vartheta^2, \vartheta^1, \vartheta^2)$ by g, we have

$$f(x) = g(x + \vartheta^1 \vartheta^2, \vartheta^1, \vartheta^2)$$
$$= g_0(x) + g_0'(x)\vartheta^1 \vartheta^2 + g_1(x)\vartheta^1 \vartheta^2.$$

Here, we have used the rule (1.5.20) for the Taylor expansion for a function g_0 of the even coordinates, and we then need to add a counter-term $g_1(x) = -g_0'(x)$ in order to compensate for the $\vartheta^1 \vartheta^2$ term from the Taylor expansion of g_0. Note that here we work over a trivial base S.

If we are Taylor-expanding functions as explained, then if M_0 is an ordinary manifold, that is, a supermanifold with odd dimension 0, and if we consider a map $f_S : M_0 \times S \to N$, then the odd dimension of S determines the maximal degree occurring in that expansion of f_S. In the physics literature, one expresses this by fixing the number N of Grassmann generators. In the present framework, this corresponds to the odd dimension of S.

One should also note that a super vector space $W = W_0 \oplus W_1$ is not a supermanifold, unless the odd part W_1 is trivial. If the even and odd part have dimensions m and n, resp., then W has the underlying structure of an $m + n$-dimensional ordinary vector space, whereas the ordinary manifold M_r underlying an $(m|n)$-dimensional supermanifold is only m-dimensional. Of course, one can canonically construct a supermanifold from a super vector space, but as such, the two structures of a super vector space and of a supermanifold are different.

A super Lie group is a supermanifold that is functorially characterized by the property that for all supermanifolds S, $\text{Hom}(S, M)$ is a group such that the group operations are smooth morphisms of supermanifolds. It can be obtained by exponentiation from a super Lie algebra. More precisely, however, for that exponentiation, we also need to be able to multiply the elements of the super Lie algebra by the odd variables ϑ^j, that is, on the super Lie algebra, we also need the structure of a left supermodule over the algebra spanned by the ϑ^j.[18] For a super Lie group H, we can consider left multiplication by an element h

$$L_h : H \to H, \qquad L_h(k) = hk \quad \text{for } k \in H. \tag{1.5.30}$$

This induces a map $(L_h)_\star$ on the vector fields on H, given by

$$((L_h)_\star X)F := X(F \circ L_h) \quad \text{for functions } F. \tag{1.5.31}$$

When $(L_h)_\star X = X$ for all $h \in H$, the vector field is called left-invariant. The left-invariant vector fields then span a super Lie algebra (with the graded commutator of vector fields as the bracket) that is also a super module over the odd variables.

To see the principle, we consider $\mathbb{R}^{1|1}$ with coordinates t, ϑ. This space carries a super Lie group structure given by

$$(t^1, \vartheta^1)(t^2, \vartheta^2) = (t^1 + t^2 + \vartheta^1 \vartheta^2, \vartheta^1 + \vartheta^2). \tag{1.5.32}$$

The translation in the t-direction is generated by the vector field

$$\partial_t \left(:= \frac{\partial}{\partial t} \right), \tag{1.5.33}$$

the one in the ϑ-direction by

$$D := \partial_\vartheta - \vartheta \partial_t. \tag{1.5.34}$$

We note that D does not induce a morphism in our sense as it changes the parity. We have the relation

$$[D, D] = 2D^2 = -2\partial_t. \tag{1.5.35}$$

(D, ∂_t) constitute a basis of the left invariant vector fields on the super Lie group, while $(Q := \partial_\vartheta + \vartheta \partial_t, \partial_t)$ is a basis for the right invariant ones. We shall meet these vector fields when we consider supersymmetry transformations. The important point is that they generate diffeomorphisms of the superspace $\mathbb{R}^{1|1}$.

Remark The treatment of supermanifolds presented here has been developed by Leites [76], Manin [79], Bernstein, Deligne and Morgan [24], and Freed [39, 40]. Another reference is [102]. The comprehensive presentation of the subject is [11]. The superdiffeomorphism group is investigated in [94].

1.5.3 Super Riemann Surfaces

As an example, we now consider super Riemann surfaces (SRSs). While above, we have defined supermanifolds over $\mathbb{R}$, it is straightforward to develop the same constructions over $\mathbb{C}$. An SRS then has one commuting complex coordinate z and

---

[18]We only have a supermodule instead of a super vector space because that algebra is only a ring, but not a field.

one anticommuting one ϑ. In addition, the coordinate transformations are required
to be superconformal. To explain this, we start with the coordinate transformation
formula for a single supercomplex manifold M of complex dimension $(1|1)$, which
has to be even, that is, of the form

$$\tilde{z} = f(z),$$
$$\tilde{\vartheta} = \vartheta h(z) \tag{1.5.36}$$

with holomorphic functions f and h where f is required to have a nonvanishing derivative, that is, to be conformal. The structure sheaf is thus of the form
$\mathcal{O}_M = \mathcal{O}_{M,0} \oplus \mathcal{O}_{M,1}$ where $\mathcal{O}_{M,0}$ is the sheaf of holomorphic functions on the under-
lying Riemann surface M_r and $\mathcal{O}_{M,1}$ is a sheaf of locally free modules of rank $0|1$
over $\mathcal{O}_{M,0}$. Up to a change of parity, this then defines a line bundle L over M_r, and
conversely, given such a line bundle L over M_r, changing the parity of its sections
from even to odd then defines the structure sheaf of supercomplex manifold of di-
mension $(1|1)$. Thus, such $(1|1)$-dimensional supercomplex manifolds and ordinary
Riemann surfaces with a line bundle L stand in bijective correspondence. When we
look at families of such supercomplex manifolds, however, we may also take base
spaces with odd directions, and we have the more general transformation formula

$$\tilde{z} = f(z) + \vartheta k(z),$$
$$\tilde{\vartheta} = g(z) + \vartheta h(z) \tag{1.5.37}$$

with holomorphic functions f, k, g, h and f again conformal.

In order to define a super Riemann surface, we require in addition that the structure be superconformal. This means the following: We look at the derivative operators ∂_z and $\tau := \partial_\vartheta + \vartheta \partial_z$; they satisfy

$$\frac{1}{2}[\tau, \tau] = \tau^2 = \partial_z. \tag{1.5.38}$$

We have the transformation rule

$$\tau = (\tau \tilde{\vartheta})\tilde{\tau} + (\tau \tilde{z} - \tilde{\vartheta} \tau \tilde{\vartheta})\tilde{\tau}^2. \tag{1.5.39}$$

(To see this, one computes

$$\partial_z = (f_z + \vartheta k_z)\partial_{\tilde{z}} + (g_z + \vartheta h_z)\partial_{\tilde{\vartheta}}, \tag{1.5.40}$$

$$\partial_\vartheta = h\partial_{\tilde{\vartheta}} + k\partial_{\tilde{z}}, \tag{1.5.41}$$

$$\tau \tilde{\vartheta} = h + \vartheta g_z, \tag{1.5.42}$$

$$\tau \tilde{z} = k + \vartheta f_z \tag{1.5.43}$$

from which

$$\tau = \partial_\vartheta + \vartheta \partial_z = (h + \vartheta g_z)\partial_{\tilde{\vartheta}} + (k + \vartheta f_z)\partial_{\tilde{z}}$$
$$= (h + \vartheta g_z)(\partial_{\tilde{\vartheta}} + \tilde{\vartheta} \partial_{\tilde{z}}) - (g + \vartheta h)(h + \vartheta g_z)\partial_{\tilde{z}} + (k + \vartheta f_z)\partial_{\tilde{z}}$$
$$= (\tau \tilde{\vartheta})\tilde{\tau} + (\tau \tilde{z} - \tilde{\vartheta} \tau \tilde{\vartheta})\tilde{\tau}^2 \tag{1.5.44}$$

which is the required formula.) In the same manner as for an ordinary Riemann surface, that is, one with transition functions $\tilde{z} = f(z)$, the holomorphicity of f implies that ∂_z is a multiple of $\partial_{\tilde{z}}$, $\partial_z = \partial_z f \partial_{\tilde{z}}$. We now require for an SRS that τ transforms homogeneously, that is, τ is a multiple of $\tilde{\tau}$. In view of (1.5.39), for a family, this leads to the transformation law

$$\tilde{z} = f(z) + \vartheta g(z)h(z),$$
$$\tilde{\vartheta} = g(z) + \vartheta h(z)$$

(1.5.45)

with

$$h^2(z) = \partial_z f(z) + g(z)\partial_z g(z).$$

(1.5.46)

Here, $f(z)$ is a commuting holomorphic function with $\frac{\partial}{\partial z} f \neq 0$, i.e., f is conformal, and $g(z)$ is an anticommuting one.

These transformations then leave the line element $dz + \vartheta d\vartheta$ invariant up to conformal scaling. (The conformal factor is $\frac{\partial}{\partial z} f(z) + g(z)\frac{\partial}{\partial z} g(z)$, and one has to use (1.5.46).)

Given a single SRS Σ, we can put all the $g = 0$ and obtain the transformation rules

$$\tilde{z} = f(z),$$
$$\tilde{\vartheta} = \vartheta h(z)$$

(1.5.47)

with $h^2(z) = \partial_z f(z)$. The holomorphic transformation functions f of z define an ordinary Riemann surface Σ_r, but the transformations of the odd coordinate ϑ additionally require the choice of a square root $h(z)$ of $\frac{\partial}{\partial z} f(z)$. In other words, they determine a spin structure on Σ_r. If p is the genus of Σ_r, we have 2^{2p} different spin structures on Σ_r. In particular, we see that the super Teichmüller space of super Riemann surfaces of genus p has at least 2^{2p} components (this does not hold for the super moduli space, because modular transformations can mix the spin structures). By the Riemann–Roch theorem (stated in 3 of Sect. 1.4.2 and recalled below), the number of even moduli (over $\mathbb{C}$) minus the number of conformal transformations of Σ_r is $3p - 3$ while the number of odd moduli minus the number of odd superconformal transformations is $2p - 2$. (The even moduli here can be identified with sections of K^2, where K is the canonical bundle of the underlying Riemann surface, while the odd ones correspond to sections of $K^{3/2}$. The Riemann–Roch theorem says that the space of sections of a line bundle L over a Riemann surface Σ of genus p has dimension

$$h^0(\Sigma, L) = \deg L - p + 1 + h^0(\Sigma, K \otimes L^{-1})$$

(1.5.48)

and the degree of the canonical bundle is $2p - 2$.)

On a sphere, we have no nontrivial spin structures and no super moduli, but, in agreement with the Riemann–Roch theorem, the superconformal transformations are of the form $f(z) = \frac{az+b}{cz+d}$ (with the normalization $ad - bc = 1$), $g(z) = \frac{\gamma z + \delta}{cz+d}$, that is, 3 even and 2 odd parameters.

More generally, on the supersphere, when, instead of (1.5.47), we allow for the general type of coordinate transformations (1.5.37), we obtain the orthosymplectic group $OSp(1/2)$:

$$T = \begin{pmatrix} a & b & \alpha \\ c & d & \beta \\ \gamma & \delta & t \end{pmatrix},$$

a, b, c, d, t even (commuting), $\alpha, \beta, \gamma, \delta$ odd (anticommuting).

$$T^{st} K T = K \quad (T^{st} \text{ supertransposed}),$$

with the orthosymplectic form

$$K = \begin{pmatrix} 0 & 1 & 0 \\ -1 & 0 & 0 \\ 0 & 0 & 1 \end{pmatrix}.$$

The transformation

$$z \mapsto \frac{a\,z + b + \alpha\vartheta}{cz + d + \beta\vartheta}, \qquad \vartheta \mapsto \frac{\gamma z + \delta + t\vartheta}{cz + d + \beta\vartheta}$$

leaves the line element $dz + \vartheta\, d\vartheta$ invariant up to conformal scaling.

In that case, just to see some formulae,

$$dz \mapsto \frac{dz}{(cz + d + \beta\vartheta)^2},$$

$$z_{12} = z_1 - z_2 - \vartheta_1\vartheta_2 \mapsto \frac{z_{12}}{(cz_1 + d + \beta\vartheta_1)(cz_2 + d + \beta\vartheta_2)},$$

$$dz \wedge d\vartheta \mapsto \frac{dz \wedge d\vartheta}{cz + d + \beta\vartheta}.$$

Obviously, this extends the operation of $Sl(2, \mathbb{C})$ to the super case.

The supersphere can be covered by two coordinate patches, with transition

$$\tilde{z} = \frac{1}{z}, \qquad \tilde{\vartheta} = \frac{i\vartheta}{z}.$$

(Cf. (1.5.45): Here, $h = \sqrt{\frac{\partial}{\partial z} f}$.)

Genus 1 is next. A torus with a spin structure is described by the rigid super conformal transformation

$$(z, \vartheta) \cong (z + 1, \eta_1\vartheta) \cong (z + \tau, \eta_2\vartheta),$$

where τ is taken from the usual period domain, and the $\eta_i = \pm 1$ determine the spin structure. Since the only holomorphic functions on a torus are the constants, we obtain a nontrivial supermodulus ν only in the case of a trivial spin structure, that is, $\eta_1 = 1 = \eta_2$. In that case, the periodicities are

$$(z, \vartheta) \cong (z + 1, \vartheta) \cong (z + \tau + \vartheta\nu, \vartheta + \nu).$$

In agreement with Riemann–Roch, we then also have the odd superconformal transformation given infinitesimally as

$$(z, \vartheta) \mapsto (z + \vartheta\varepsilon, \vartheta + \varepsilon).$$

In particular, we see that there can exist nontrivial odd moduli, and the supermoduli space is bigger than just the moduli space of ordinary Riemann surfaces with spin structures. We observe, however, that $\varepsilon \mapsto -\varepsilon$ is a superconformal transformation, and so the supertori corresponding to ε and $-\varepsilon$ are equivalent. Thus, the corresponding component of the supermoduli space is a $\mathbb{Z}_2$ super orbifold, with a singularity at $\varepsilon = 0$.

When we look at functions on this super torus, we obtain the periodicity condition

$$f(z, \vartheta) = f(z + \tau + \vartheta\nu, \vartheta + \nu),$$

that is, after Taylor expanding,

$$f_0(z) + f_1(z)\vartheta = f_0(z + \tau) + f_0'(z + \tau)\vartheta\nu + f_1(z + \tau)(\vartheta + \nu)$$

which implies that f_0' vanishes when $\nu \neq 0$, that is, f_0 is constant (over a trivial base S again). f_1 is less trivial. The situation becomes richer when we look at mappings between two such supertori, with moduli (τ, ν) and $(\tilde{\tau}, \tilde{\nu})$, resp. We then expand to obtain

$$f_0(z) + \tilde{\tau} + f_1(z)\vartheta\tilde{\nu} = f_0(z + \tau) + f_0'(z + \tau)\vartheta\nu$$

and

$$f_1(z)\vartheta + \tilde{\nu} = f_1(z + \tau)(\vartheta + \nu).$$

The first equation expresses f_0' in terms of f_1 or conversely, while the second one restricts f_1. However, we should be careful here as f_0 need not be holomorphic, and so f_0' stands for a (2×2)-matrix.

Remark For a treatment of super Riemann surfaces as needed for superstring theory, we refer to Crane and Rabin [21, 89] and Polchinski [88]. A general mathematical perspective is developed by Leites and his coauthors in [32]. A very lucid discussion, which we have also partly utilized here, can be found in [93].

We should note that the above definition is not the only possible for an SRS. In fact, there are several superextensions of the conformal algebra, and each of them could be taken as the basis for the definition of an SRS. The one used here corresponds to the superconformal algebra $\mathfrak{k}^L(1|1)$ and yields the $N = 1$ worldsheets of superstring theory and $2D$ supergravity.

1.5.4 Super Minkowski Space

Now assume that we have a vector space V with a quadratic form Q and a representation of the Clifford algebra $Cl(Q)$ for which a symmetric equivariant morphism Γ

as in (1.3.25) exists. Following the presentation in [22], we may then construct an object that captures deeper aspects of the physical concept of supersymmetry than just a supermanifold, namely a space that incorporates the symmetry between vectors and spinors as representations of bosons and fermions, resp. For that purpose, we consider the vector space V as the Lie algebra of its translations, and construct the super Lie algebra

$$\mathfrak{l} := V \oplus S^\star \qquad\qquad (1.5.49)$$

and the bracket $[.,.]$. This bracket is trivial on V (that is, V is central) and is given by

$$[s, t] = -2\Gamma(s, t) \in V \qquad\qquad (1.5.50)$$

on $S^\star$. Super Minkowski space M is then defined as the supermanifold underlying the Lie group $\exp(\mathfrak{l})$; its reduced space is thus given by the affine space V, and its odd directions are given by $S^\star$.

In the Minkowski case, the super Lie algebra (1.5.49) leads to the super Poincaré algebra

$$(V \oplus \mathfrak{so}(V)) \oplus S^\star. \qquad\qquad (1.5.51)$$

Chapter 2
Physics

2.1 Classical and Quantum Physics

2.1.1 Introduction

In this section, we will describe some important principles at a heuristic level. We hope this will be useful as a guide to some of the sequel which is more formal, but whenever the meaning of this section appears unclear, the reader should proceed to the more formal treatment below. There are many textbooks available on the mathematical aspects of quantum mechanics, for instance [53].

Classically, a particle is represented as or described by a point in some state space M. It moves in time along some trajectory $x(t)$ that is a solution of a system of second-order ODEs (a dot denoting a derivative with respect to time t),

$$\ddot{x}(t) = f(x, \dot{x}) \tag{2.1.1}$$

that is derived from an action principle. This principle consists in minimizing the Lagrangian action

$$S(x) := \int F(x(t), \dot{x}(t)) \, dt, \tag{2.1.2}$$

the integral w.r.t. time over some Lagrangian that is a function of x and its first temporal derivative.[1] As will be discussed in more detail in Sect. 2.3.1 below, a minimizing $x(t)$ satisfies the corresponding Euler–Lagrange equations

$$\frac{d}{dt}\frac{dF}{d\dot{x}} - \frac{dF}{dx} = 0. \tag{2.1.3}$$

Here, the space M is d-dimensional, and in local coordinates $x = (x^1, \ldots, x^d)$. Alternatively, one may utilize the $2d$-dimensional phase space N with coordinates $(x^1, \ldots, x^d, x^{d+1} = \dot{x}^1, \ldots, x^{2d} = \dot{x}^d)$.

$\frac{dF}{d\dot{x}}$ stands for the covector of partial derivatives $(\frac{\partial F}{\partial \dot{x}^1}, \ldots, \frac{\partial F}{\partial \dot{x}^d})$. When one introduces this covector as a new variable, that is, puts

$$p := \frac{dF}{d\dot{x}}, \quad \text{i.e.,} \quad p_j := \frac{\partial F}{\partial \dot{x}^j} \tag{2.1.4}$$

---

[1] We consider here only the autonomous case; in the non-autonomous case, the density may also explicitly depend on t, $F(t, x(t), \dot{x}(t))$, and not only implicitly through its dependence on $x(t)$ and $\dot{x}(t)$.

J. Jost, *Geometry and Physics*,
DOI 10.1007/978-3-642-00541-1_2, © Springer-Verlag Berlin Heidelberg 2009

one arrives at the Hamiltonian formulation. This involves the Hamiltonian

$$H(p, x) := \dot{x}\frac{dF}{d\dot{x}} - F \tag{2.1.5}$$

where $\dot{x}\frac{dF}{d\dot{x}} = \sum_j \dot{x}^j \frac{\partial F}{\partial \dot{x}^j} = \sum_j \dot{x}^j p_j$. A solution is then obtained from the Hamilton equations

$$\dot{p} = -\frac{dH}{dx}, \qquad \dot{x} = \frac{dH}{dp}. \tag{2.1.6}$$

The Hamiltonian formalism singles out time and is therefore not relativistically invariant. Consequently, in our treatment of QFT, we shall mainly employ the Lagrangian formalism.

The standard example is

$$F = \frac{m}{2}|\dot{x}|^2 - V(x), \tag{2.1.7}$$

where m is the mass of the particle and V the potential. The Euler–Lagrange equations (2.1.3) are then

$$m\ddot{x} = -\frac{dV}{dx} \tag{2.1.8}$$

(in components: $m\ddot{x}^i = -\frac{\partial V}{\partial x^i}$). The Hamiltonian is then

$$H = \frac{m}{2}|\dot{x}|^2 + V(x) = \frac{p^2}{2m} + V(x), \tag{2.1.9}$$

and (2.1.6) becomes

$$\dot{p} = -\frac{dV}{dx}, \qquad \dot{x} = \frac{p}{m}. \tag{2.1.10}$$

For a solution $(x(t), p(t))$ of (2.1.6), we can then also compute the time evolution of any function $A(x, p)$ via

$$\frac{dA}{dt} = \frac{\partial A}{\partial x^i}\dot{x}^i + \frac{\partial A}{\partial p^i}\dot{p}^i = \frac{\partial A}{\partial x^i}\frac{\partial H}{\partial p_i} - \frac{\partial A}{\partial p_i}\frac{\partial H}{\partial x^i} =: \{A, H\}, \tag{2.1.11}$$

where the last expression is called the Poisson bracket. It satisfies all the properties of a Lie bracket, as well as the canonical relations (Heisenberg commutation relations)

$$\{x^i, x^j\} = 0 = \{p_i, p_j\} \quad \text{and} \quad \{x^i, p_j\} = \delta^i_j. \tag{2.1.12}$$

Equation (2.1.11) is obviously a generalization of (2.1.6) (in the sense that $\dot{x}^i = \{x^i, H\}$, $\dot{p}_j = \{p_j, H\}$), and it also tells us that conserved quantities, that is, time-independent quantities, are precisely those whose Poisson bracket with the Hamiltonian H vanishes.

Quantum mechanics was discovered by Heisenberg and developed with Born and Jordan as the description of quantum theory through the correspondence with classical mechanics via matrix algebra. We now describe this, employing more modern terminology, of course. Quantum mechanically, in place of a point x in M, we have a probability distribution $|\phi(x)|^2$ derived from a function $\phi : M \to \mathbb{C}$ with

$$\|\phi\|_{L^2}^2 \left(= \int_M |\phi|^2(x) dvol(x) \right) = 1. \tag{2.1.13}$$

$|\phi(x)|^2$ can thus be interpreted as the probability density for finding the particle under consideration at the point x. Here, for the L^2-norm, we need a volume form $dvol$ on M; that volume form could come from a Riemannian metric. The classical case is recovered as the limit where this probability distribution becomes concentrated at a single point, that is, a delta function(al). In quantum mechanics, the observables are self-adjoint operators on the Hilbert space $\mathcal{H} := L^2(M, \mathbb{C})$. As self-adjoint operators, they have a purely real spectrum. The eigenvalues corresponding to eigenstates of such an operator then represent sharp observations. These operators, however, are typically unbounded which leads to certain mathematical difficulties, as will be described in more detail in Sect. 2.1.3 below.

In the formalism of canonical quantization, the momentum p_j becomes the operator $\frac{\hbar}{i} \frac{\partial}{\partial x^j}$. The total energy, the Hamilton function above, thus also becomes an operator, the Hamiltonian operator H, and the state ϕ evolves in time t according to the Schrödinger equation

$$i\hbar \frac{\partial \phi(x,t)}{\partial t} = H\phi(x,t). \tag{2.1.14}$$

For the Lagrangian (2.1.7), the Schrödinger equation (2.1.14) becomes

$$i\hbar \frac{\partial \phi(x,t)}{\partial t} = -\frac{\hbar^2}{2m} \Delta \phi(x,t) + V(x)\phi(x,t). \tag{2.1.15}$$

The ansatz $\phi(x,t) = \phi(x) \exp(-\frac{i}{\hbar} Et)$ of separated variables leads to

$$-\frac{\hbar^2}{2m} \Delta \phi(x) + V(x)\phi(x) = E\phi(x), \tag{2.1.16}$$

the time-independent Schrödinger equation.

We can arrive at (2.1.15) from the ansatz of representing $\phi(x,t)$ as a wave:

$$\phi(x,t) = \frac{1}{2\pi^{3/2}} \exp \frac{i}{\hbar} (p_\nu x^\nu - Et) =: \langle x, t | p, E \rangle, \tag{2.1.17}$$

where we have already introduced Dirac's notation to be explained below. Then

$$\frac{\partial}{\partial x^\nu} \phi(x,t) = \frac{i}{\hbar} p_j \phi(x,t), \tag{2.1.18}$$

$$\frac{\partial}{\partial t}\phi(x,t) = -\frac{i}{\hbar}E\phi(x,t). \tag{2.1.19}$$

On the basis of this computation, we then put

$$\frac{\partial}{\partial x^\nu} = \frac{i}{\hbar}p_j, \tag{2.1.20}$$

$$\frac{\partial}{\partial t} = -\frac{i}{\hbar}E. \tag{2.1.21}$$

For the Hamiltonian (2.1.9), we are then naturally led to (2.1.15), and with the ansatz $\phi(x,t) = \phi(x)\exp(-\frac{i}{\hbar}Et)$ at (2.1.16).

Remark We use here the so-called Schrödinger picture where the states ϕ are evolving in time. In the complementary Heisenberg picture, instead the observables, represented as self-adjoint operators A, evolve according to

$$i\hbar\frac{dA}{dt} = [A, H], \tag{2.1.22}$$

in analogy to (2.1.11), see (2.1.92), (2.1.93).

In the quantum mechanical view, the field $\phi : M \to \mathbb{C}$ is obtained from the quantization of a point particle. There is, however, another interpretation of ϕ that turns out to be more fruitful for our purposes. Namely, we can view ϕ also as a classical field on M. It then need no longer satisfy the normalization $\|\phi\|_{L^2} = 1$. Also, it need no longer take its values in $\mathbb{C}$ only, but it can also assume values in the fibers of some vector or principal bundle or some manifold. As a classical situation, it can then be quantized again, and one then speaks of a second or field quantization. The analog of the Schrödinger equation is then a PDE on some function space, that is, a PDE with infinitely many variables.

There is an important generalization of this picture: When the particle possesses some internal symmetry, described by some Lie group G, the space $\mathbb{C}$ gets replaced by a (Hermitian) vector space that carries a (unitary) representation of G. Thus, a particle is described by some $\psi \in L^2(M, V)$, again of norm 1, so that $\|\psi\|^2$ (where $\|.\|$ is the Hermitian norm) can again be interpreted as a probability density. The vector space V enters here in order to distinguish different states that are not G-invariant, as G leaves the space V invariant, but not the individual elements of V. This is needed because not all physical forces will be G-invariant. An example is the electron with its spin. Since there are only two possible values of the spin, here the vector space is finite, $\mathbb{Z}_2$, and the corresponding Hilbert space is finite-dimensional, $\mathbb{C}^2$. Quantum electrodynamics (QED) then couples the Maxwell equation with the Dirac equation for the electron spin on a relativistic space time. The standard model of elementary particle physics interprets the observed multitude of particles through symmetry breaking from some encompassing Lie group $\bar{G}$ that contains all the symmetry groups of the individual particles. Of course, we shall explain this in more detail below. A quick and useful introduction to the topics of this section can be found in [90].

2.1.2 Gaussian Integrals and Formal Computations

Before proceeding with quantum physics, we introduce a basic formal tool, Gaussian integrals, that serve as a heuristic transmission line from finite-dimensional exponential integrals to infinite-dimensional functional integrals.

We start with the bosonic case. Let A be a symmetric $n \times n$-matrix with eigenvalues

$$\lambda_i > 0, \tag{2.1.23}$$

and let b be a vector.

The Gaussian integral $(x = (x^1, \dots, x^n))$ is

$$I(A) := \int \exp\left(-\frac{1}{2} x^t A x\right) dx^1 \cdots dx^n = \left(\frac{(2\pi)^n}{\det A}\right)^{\frac{1}{2}} \tag{2.1.24}$$

with $\det A = \prod_{i=1}^{n} \lambda_i$, as follows easily by diagonalizing A. A formal extension of this formula to infinite dimensions is often based on expressing the determinant of A in terms of a zeta function; we define the zeta function of the operator A as

$$\zeta_A(s) := \sum_{k=1}^{n} \frac{1}{\lambda_k^s}, \quad \text{for } s \in \mathbb{C}. \tag{2.1.25}$$

Since $\lambda_k^{-s} = e^{-s \log s}$, we obtain for the derivative of the zeta function

$$\zeta_A'(s) = -\sum_{k=1}^{n} \frac{\log \lambda_k}{\lambda_k^s}. \tag{2.1.26}$$

Therefore, we can express the determinant of A in terms of the derivative of the zeta function at 0:

$$\det A = \prod_{i=1}^{n} \lambda_i = e^{-\zeta_A'(0)}. \tag{2.1.27}$$

The general Gaussian integral

$$I(A, b) := \int dx^1 \cdots dx^n \exp\left(-\frac{1}{2} x^t A x + b^t x\right) \tag{2.1.28}$$

is reduced to this case by putting

$$x := A^{-1}b + y$$

(note that $x_0 := A^{-1}b$ minimizes the quadratic form $\frac{1}{2}x^t Ax - b^t x$). Namely, we obtain

$$
\begin{aligned}
I(A,b) &= \exp\left(\frac{1}{2}b^t A^{-1}b\right) \int \exp\left(-\frac{1}{2}y^t A\, y\right) dy^1 \cdots dy^n \\
&= \exp\left(\frac{1}{2}b^t A^{-1}b\right)\left(\frac{(2\pi)^n}{\det A}\right)^{\frac{1}{2}} \\
&= (2\pi)^{n/2}\exp\left(\frac{1}{2}b^t A^{-1}b\right)e^{\frac{1}{2}\zeta_A'(0)},
\end{aligned}
\tag{2.1.29}
$$

using (2.1.27) for the last line.

In many cases, the vector b has an auxiliary or dummy role. Namely, we wish to compute moments

$$
\begin{aligned}
\langle x^{i_1}\cdots x^{i_m}\rangle &:= \frac{\int x^{i_1}\cdots x^{i_m}\exp(-\frac{1}{2}x^t Ax)\,dx^1\cdots dx^n}{\int \exp(-\frac{1}{2}x^t Ax)\,dx^1\cdots dx^n} \\
&= \frac{1}{I(A)}\frac{\partial}{\partial b^{i_1}}\cdots\frac{\partial}{\partial b^{i_m}}I(A,b)|_{b=0}.
\end{aligned}
\tag{2.1.30}
$$

In particular, the second-order moment or propagator is

$$
\langle x^i x^j\rangle = \left(A^{-1}\right)_{ij}.
\tag{2.1.31}
$$

When m is odd, the moment (2.1.30) vanishes because the (quadratic) exponential is even at $b=0$.

For even m, we have Wick's theorem

$$
\langle x^{i_1}\cdots x^{i_m}\rangle = \sum_{\substack{\text{all possible}\\ \text{pairings of}\\ (i_1,\dots,i_m)}} \left(A^{-1}\right)_{i_{p_1}i_{p_2}}\cdots\left(A^{-1}\right)_{i_{p_{m-1}}i_{p_m}}
\tag{2.1.32}
$$

as follows directly from (2.1.29) and (2.1.30).

As a preparation for the functional integrals to follow, we now wish to consider x^i as an operator on the finite-dimensional Hilbert space $E^n = \mathbb{R}^n$ with its Euclidean product. We then have the matrix elements

$$
\begin{aligned}
&\langle e_i|x^{i_1}\cdots x^{i_m}|e_j\rangle \\
&= \frac{\int x^{i_1}\cdots x^{i_m}\delta(x^i-1)\delta(x^j-1)\exp(-\frac{1}{2}x^t Ax)\,dx^1\cdots dx^n}{\int \exp(-\frac{1}{2}x^t Ax)\,dx^1\cdots dx^n}.
\end{aligned}
$$

Instead of the δ-functions, one can then also make arbitrary insertions into the functional integral, that is, functions of x.

In the Grassmann case, we start with a Grassmann algebra generated by $\eta^1, \ldots, \eta^n, \bar{\eta}^1, \ldots, \bar{\eta}^n$

$$J_0(A) := \int d\eta^1 \, d\bar{\eta}^1 \cdots d\eta^n \, d\bar{\eta}^n \, \exp\left(-\bar{\eta}^t \, A\eta\right)$$

$$= \int d\eta^1 \, d\bar{\eta}^1 \cdots d\eta^n \, d\bar{\eta}^n \prod_{i=1}^{n}\prod_{j=1}^{n}\left(1 - \bar{\eta}^i A_{ij}\eta^j\right)$$

$$= \sum_{\text{permutations } p} \text{sign}(p) A_{1p(1)} \cdots A_{np(n)}$$

$$= \det A. \tag{2.1.33}$$

We next compute, for a Grassmann algebra generated by $\vartheta^1, \ldots, \vartheta^{2n}$,

$$J(A) = \int d\vartheta^1 \cdots d\vartheta^{2n} \, \exp\left(-\frac{1}{2}\vartheta^t A\vartheta\right). \tag{2.1.34}$$

We may assume that A is antisymmetric, as the symmetric terms cancel because the ϑ's anticommute:

$$J(A) = \int d\vartheta^1 \cdots d\vartheta^{2n} \prod_{i<j}\left(1 - \vartheta^i A_{ij}\vartheta^j\right)$$

$$= \sum_{\substack{\text{permutations } p \\ \text{with } p(2i-1)<p(2i), \\ p(2i-1)<p(2i+1) \\ \text{for } i=1,\ldots,n, \text{ or } n-1, \text{ resp.}}} \text{sign}(p) A_{p(1)p(2)} A_{p(3)p(4)} \cdots A_{p(n-1)p(n)}$$

$$=: \text{Pf}(A) \quad \text{(Pfaffian)}. \tag{2.1.35}$$

We have

$$J^2(A) = \int d\vartheta^1 \cdots d\vartheta^{2n} \, d\vartheta'^1 \cdots \vartheta'^{2n} \, \exp\left(-\frac{1}{2}\left(\vartheta^t A\vartheta + \vartheta''^t A\vartheta'\right)\right). \tag{2.1.36}$$

The coordinate transformation

$$\eta^k := \frac{1}{\sqrt{2}}\left(\vartheta^k + i\vartheta'^k\right),$$

$$\bar{\eta}^k := \frac{1}{\sqrt{2}}\left(\vartheta^k - i\vartheta'^k\right),$$

has the Jacobian $(-1)^n$ and satisfies $\vartheta^i \vartheta^j + \vartheta'^i \vartheta'^j = \bar{\eta}^i \eta^j - \bar{\eta}^j \eta^i$.

Using the antisymmetry of A, we obtain

$$J^2(A) = \int d\eta^1 \, d\bar{\eta}^1 \cdots d\eta^n \, d\bar{\eta}^n \exp(-\bar{\eta}^t A \eta)$$

$$= \det A \tag{2.1.37}$$

from (2.1.33). From (2.1.34), (2.1.35), (2.1.37) we see

$$\mathrm{Pf}^2(A) = \det A, \tag{2.1.38}$$

that is,

$$J(A) = (\det A)^{\frac{1}{2}}. \tag{2.1.39}$$

As in the ordinary case, we also have

$$J(A,b) = \int d\vartheta^1 \cdots d\vartheta^{2n} \exp\left(-\frac{1}{2}\vartheta^t A \vartheta + b^t \vartheta\right)$$

$$= J(A) \exp\left(\frac{1}{2}b^t A^{-1} b\right) \tag{2.1.40}$$

and likewise

$$\langle \vartheta^i \vartheta^j \rangle = \frac{1}{J(A)} \frac{\partial}{\partial b_j} \frac{\partial}{\partial b_i} J(A,b)_{|b=0} = (A^{-1})_{ij}.$$

Another formal tool that is useful in this context are formal computations with Dirac functions. In the physics literature, linear functionals on space of functions are systematically expressed by their integral kernels. Thus, the evaluation of a function φ at a point y

$$\varphi \mapsto \varphi(y) \tag{2.1.41}$$

is written in terms of the Dirac δ-functional

$$\varphi(y) = \int dz \varphi(z)\delta(z-y) = \delta_y(\varphi). \tag{2.1.42}$$

If we change the variable $z = f(w)$, this becomes

$$\varphi(y) = \int dw \left|\det \frac{\partial f}{\partial w}\right| \varphi(f(w))\delta(f(w) - y). \tag{2.1.43}$$

This formula is useful for calculating $\varphi(f(w))$ at $f(w) = y$, without having to solve the latter equation explicitly.

In analogy to (2.1.42), we also have

$$\frac{\partial}{\partial y^j}\varphi(y) = \int dz \frac{\partial}{\partial y^j}\varphi(z)\delta(z-y) = -\int dz\varphi(z)\frac{\partial}{\partial y^j}\delta(z-y) \tag{2.1.44}$$

if we formally integrate by parts. Thus, we can define the derivative $\frac{\partial}{\partial y^j}\delta(z-y)$ of the delta function $\delta(z-y)$ as the functional

$$\varphi \mapsto -\frac{\partial}{\partial y^j}\varphi(y). \qquad (2.1.45)$$

We now consider a functional Φ

$$\varphi \mapsto \Phi(\varphi)$$

defined on some Banach or Fréchet space $\mathcal{B}$. The Gateaux derivative $\delta\Phi$ in the direction $\delta\varphi$ is defined as

$$\delta\Phi(\varphi)(\delta\varphi) := \frac{\delta\Phi(\varphi)}{\delta\varphi} := \lim_{t\to 0}\frac{1}{t}(\Phi(\varphi+t\delta\varphi)-\Phi(\varphi)), \qquad (2.1.46)$$

provided this limit exists. Here, $\delta\varphi$ is, of course, also assumed to be in $\mathcal{B}$. When the limit in (2.1.46) exists uniformly for all variations $\delta\varphi$ in some neighborhood of 0, we speak of a Fréchet derivative. Some formal examples:

$$\delta\varphi(f)^n(\delta\varphi) = n\delta\varphi(f)\varphi(f)^{n-1}, \qquad (2.1.47)$$

$$\delta e^{\varphi(f)}(\delta\varphi) = \delta\varphi(f)e^{\varphi(f)}. \qquad (2.1.48)$$

We are usually interested in Lagrangian functionals,

$$L(u) := \int F(\xi, u(\xi), du(\xi))\, d\xi. \qquad (2.1.49)$$

L is usually defined on some Sobolev space of functions. We then have

$$\delta L(u)(\delta u) = \frac{\delta L(u)}{\delta u} = \frac{d}{ds}\int F(\xi, u(\xi)+s\delta u(\xi), d(u(\xi)+s\delta u(\xi)))\, d\xi_{|s=0}. \qquad (2.1.50)$$

The question arises as to which variations δu one may take here. One class of variations is given by the test functions, that is, the functions from the space $\mathcal{D} := C_0^\infty$ of infinitely often differentiable functions with compact support. That space is not a Banach space, but only a limit of Fréchet spaces with topology generated by the seminorms $|f|_{k,K} := \sup_{x\in K}|D^k f(x)|$ for nonnegative integers k and compact sets K. Its dual space $\mathcal{D}'$, that is, the space of continuous functionals on $\mathcal{D}$, is the space of distributions. The best-known distribution is of course the Dirac distribution already displayed in (2.1.42) above. Conceiving the Dirac functional as an element of $\mathcal{D}'$, that is, as an operation on smooth functions (in fact, continuous functions are good enough here), is the Schwartz point of view. A different point of view, which does not need topologies with unpleasant properties (for example, the implicit function theorem is very cumbersome in Fréchet spaces) and is more useful in nonlinear analysis, is the one of Friedrichs, which considers the Dirac function as a limit of smooth integral kernels with compact support that in the limit shrinks to

a point. One then, in effect, never needs to carry out any formal manipulation with
the Dirac function(al) itself, but only ones with such smooth integral kernels. The
Dirac point of view, finally, simply performs formal operations with the Dirac function. Thus, the different approaches consist in justifying, avoiding, or performing the
Dirac function computations. The Dirac approach is prominent in the physics literature, and we shall also follow that here, because we are assured that these operations
can be made mathematically rigorous by either of the other two approaches.

Having said that, we then also take functional derivatives in the direction of Dirac
functions. That means considering

$$\frac{\delta \Phi}{\delta \varphi(z)} := \lim_{t \to 0} \frac{1}{t} \left(\Phi(\varphi(y) + t\delta(y - z)) - \Phi(\varphi(y)) \right). \tag{2.1.51}$$

For (2.1.49), (2.1.50), we then have the formal relation

$$\delta L(\delta u) = \int d\xi \, \delta u(\xi) \frac{\delta L}{\delta u(\xi)}, \tag{2.1.52}$$

with the consistency relation

$$\frac{\delta L}{\delta u(z)} = \int d\xi \, \delta(z - \xi) \frac{\delta L}{\delta u(\xi)}. \tag{2.1.53}$$

Thus, $\frac{\delta L}{\delta u(z)}$ measures the response of L to a change in u supported at z.
We also have

$$\frac{\delta}{\delta \varphi(z)} \varphi(x) = \delta(x - z) \tag{2.1.54}$$

and

$$\frac{\delta}{\delta \varphi(z)} \int dx \varphi^n(x) = n\varphi^{n-1}(z). \tag{2.1.55}$$

Looking at (2.1.42), (2.1.51), (2.1.54), we see that the operations with the Dirac δ-
function are simply formal extensions of the ones with the Kronecker symbol in the
finite-dimensional case.

In the Grassmann case, the Dirac δ-function is

$$\delta(\vartheta) = \vartheta = \int d\eta \exp(\eta \vartheta) \tag{2.1.56}$$

satisfying

$$\int d\vartheta \, \delta(\vartheta) f(\vartheta) = f(0), \quad \text{for } f(\vartheta) = f(0) + a\vartheta. \tag{2.1.57}$$

For more details on the formal calculus, see [114].

2.1.3 Operators and Functional Integrals

In this section, we want to amplify the discussion of Sect. 2.1.1 and introduce path integrals. We want to investigate the time evolution of a quantized particle. This is described by a complex-valued wave function $\phi(x, t)$ whose squared norm $|\phi(x, t)|^2$ represents the probability density for finding the particle at time t at the position $x \in M$. Here, $\phi(x, t)$ is assumed to be an L^2-function of x so that the total probability can be normalized:

$$\int_M |\phi(x, t)|^2 dvol(x) = 1 \qquad (2.1.58)$$

for all t. For a measurable subset B of M, the probability for finding the particle in B at time t is then given by

$$\int_B |\phi(x, t)|^2 dvol(x). \qquad (2.1.59)$$

More abstractly, a pure state $|\psi\rangle$ of a quantum mechanical system is a one-dimensional subspace, which we then represent by a unit vector ψ, in some Hilbert space $\mathcal{H}$. The scalar product is written as $\langle \phi | \psi \rangle$; here, by duality, we may also consider $\langle \phi |$ as an element of the dual space $\mathcal{H}^*$. For a pure state ψ, we let P_ψ be the projection onto the one-dimensional subspace defined by ψ. As a projection, P_ψ is idempotent, that is, $P_\psi^2 = P_\psi$. Then

$$|\langle \phi, \psi \rangle|^2 = \langle P_\phi \psi, \psi \rangle = \text{tr } P_\phi P_\psi \qquad (2.1.60)$$

is the probability of finding the system in the state ϕ when knowing that it is in the state ψ. Let us assume that for some map T on the states of $\mathcal{H}$, we have

$$|\langle T\phi, T\psi \rangle|^2 = |\langle \phi, \psi \rangle|^2 \qquad (2.1.61)$$

for all ϕ, ψ, that is, the probabilities are unchanged by applying T to all states. By a theorem of Wigner, T can then be represented by a unitary or antiunitary operator U_T of $\mathcal{H}$, that is $T\psi = U_T \psi$ for all ψ.[2]

The observables are self-adjoint (Hermitian) operators A on $\mathcal{H}$, typically unbounded. Being self-adjoint, their spectrum is real. The state $|\psi\rangle$ then also defines an observable, the projection P_ψ. The expectation value of the observable A in the state $|\psi\rangle$ is given by

$$\langle \psi, A\psi \rangle = \text{tr } A P_\psi \qquad (2.1.62)$$

---

[2] In particular, connected groups of automorphisms G of $\mathcal{H}$ are represented by unitary transformations of $\mathcal{H}$—with the following note of caution: U_T is determined by T only up to multiplication by a factor of norm 1. Therefore, in general, we only obtain a projective representation of G, that is, we only obtain the group law $U_{gh} = c(g, h) U_g U_h$ for some scalar factor $c(g, h)$ of absolute value 1. It is, however, possible, to obtain an honest unitary representation by enlarging the group G.

(assuming that ψ is contained in the domain of definition of A). This includes (2.1.60) as a special case. When ψ is an eigenstate of A, that is,

$$A\psi = \lambda\psi \qquad (2.1.63)$$

for some (real) eigenvalue λ, then this λ is the expectation value of A in the state ψ. The variance of the probability distribution for the observations of the values of A for a system in state ψ is then

$$\langle \psi, A^2\psi \rangle - \langle \psi, A\psi \rangle^2. \qquad (2.1.64)$$

This variance vanishes iff (2.1.63) holds, that is, iff ψ is an eigenstate of A. That means that an observable A takes precise values precisely on its eigenstates.

Let G be a group, like $SO(3)$ or $SU(2)$, acting on $\mathcal{H}$ by unitary transformations. An observable A is called a scalar operator when it commutes with the action of G. Then, if ψ is an eigenstate of A with eigenvalue λ, for all $g \in G$,

$$Ag\psi = gA\psi = g\lambda\psi = \lambda g\psi. \qquad (2.1.65)$$

Thus, the space of eigenstates with eigenvalue λ is invariant under the action of G. As such an invariant subspace, it could be reducible or irreducible. In the latter case, the degeneracy of the eigenvalue λ equals the dimension of the corresponding irreducible representation of G. These dimensions are known by representation theory, see [45, 75]. If one then perturbs the operator A to an operator A' that is no longer invariant under the action of G, the multiplicity of the eigenvalue λ will decrease. This is important for understanding many experimental results. The operator A might be the Hamiltonian H_0 of a system invariant under some group G, say of spatial rotations. Its eigenvalues are the energy levels, and because of the invariance, they are degenerate. H_0 then is perturbed to $H = H_0 + H_1$ by some external magnetic field in some direction which then destroys rotational invariance. Then the energy levels, the eigenvalues of the new Hamiltonian H split up into several values. Often H_1 is small compared to H_0, and one can then approximate these energy levels by a perturbative expansion of H around H_0.

We return to the general theory. When the spectrum of A is discrete, and $|a\rangle$ runs through a complete set of orthonormal eigenvectors of A, we have the relation

$$\sum_{|a\rangle} |a\rangle\langle a| = \text{id}, \qquad (2.1.66)$$

the identity operator on $\mathcal{H}$. Applying this to $|\psi\rangle \in \mathcal{H}$ yields

$$\sum_{|a\rangle} |a\rangle\langle a|\psi\rangle = |\psi\rangle, \qquad (2.1.67)$$

which is simply the expansion of $|\psi\rangle$ in terms of a Hilbert space basis. When the eigenvalue of A corresponding to the eigenstate $|a\rangle$ is denoted by a, that is, $A|a\rangle =$

$a|a\rangle$, we have the relationships

$$A|\psi\rangle = \sum_{|a\rangle} A|a\rangle\langle a|\psi\rangle = \sum_{|a\rangle} a|a\rangle\langle a|\psi\rangle \quad \text{and} \tag{2.1.68}$$

$$\langle\phi|A|\psi\rangle = \sum_{|a\rangle} \langle\phi|A|a\rangle\langle a|\psi\rangle = \sum_{|a\rangle} a\langle\phi|a\rangle\langle a|\psi\rangle. \tag{2.1.69}$$

When the spectrum of the operator A is continuous, the sums in the preceding relations are replaced by integrals; for instance

$$\langle\phi|A|\psi\rangle = \int_a da\, a\langle\phi|a\rangle\langle a|\psi\rangle. \tag{2.1.70}$$

This is rigorously investigated in von Neumann's spectral theory of (unbounded) self-adjoint operators on Hilbert spaces. Let us briefly describe this, referring e.g. to [100, 110] for details (the knowledgeable reader may of course skip this).

A family of projections $E(a)$ $(-\infty < a < \infty)$ in a Hilbert space $\mathcal{H}$ is called a resolution of the identity iff for all $a, b \in \mathbb{R}$

1.

$$E(a)E(b) = E(\min(a,b)), \tag{2.1.71}$$

2.

$$E(-\infty) = 0, \qquad E(\infty) = Id \tag{2.1.72}$$

(here, $E(-\infty)|\phi\rangle := \lim_{a\downarrow-\infty} E(a)|\phi\rangle$, $E(\infty)|\phi\rangle := \lim_{a\uparrow\infty} E(a)|\phi\rangle$ for $|\phi\rangle \in \mathcal{H}$),

3.

$$E(a+0) = E(a) \tag{2.1.73}$$

$(E(a+0)|\phi\rangle := \lim_{b\downarrow 0} E(b)|\phi\rangle)$.

For a continuous function $f : \mathbb{R} \to \mathbb{C}$, one can then define

$$\int_{A_1}^{A_2} f(a)\, dE(a)|\phi\rangle := \lim_{\max|a_{k+1}-a_k|\to 0} \sum_k f(\alpha_k)(E(a_{k+1}) - E(a_k))|\phi\rangle \tag{2.1.74}$$

for $A_1 = a_1 < a_2 \cdots < a_n = A_2$ and $\alpha_k \in (a_k, a_{k+1}]$ (a limit of Riemann sums), and

$$\int f(a)\, dE(a)|\phi\rangle := \lim_{A_1\downarrow-\infty, A_2\uparrow\infty} \int_{A_1}^{A_2} f(a)\, dE(a)|\phi\rangle \tag{2.1.75}$$

whenever that limit exists. This is the case precisely if

$$\int |f(a)|^2\, d\|E(a)|\phi\rangle\|^2 < \infty \tag{2.1.76}$$

where $\|.\|$ is the norm in the Hilbert space $\mathcal{H}$. For such $|\phi\rangle$,

$$|\psi\rangle \to \int f(a)\, d\langle\psi|E(a)|\phi\rangle \tag{2.1.77}$$

defines a bounded linear functional on $\mathcal{H}$. In other words, we have a self-adjoint operator G with

$$\langle\psi|G|\phi\rangle = \int f(a)\,d\langle\psi|E(a)|\phi\rangle \qquad (2.1.78)$$

defined for those $|\phi\rangle$ and all $|\psi\rangle \in \mathcal{H}$. The central result is that every self-adjoint operator A on the Hilbert space $\mathcal{H}$ admits a unique spectral resolution, that is, can be uniquely written as

$$A = \int a\,dE(a), \qquad (2.1.79)$$

in the sense that

$$\langle\psi|A|\phi\rangle = \int a\,d\langle\psi|E(a)|\phi\rangle \qquad (2.1.80)$$

for all $|\phi\rangle$ in the domain of definition of A and all $|\psi\rangle \in \mathcal{H}$. This is the meaning of (2.1.70).

On this basis, one can define the functional calculus for self-adjoint operators and put

$$f(A) := \int f(a)\,dE(a) \qquad (2.1.81)$$

for a function $f : \mathbb{R} \to \mathbb{C}$ when $(E(a))$ is the spectral resolution of A. $f(A)$ is then also a self-adjoint operator. When f is an exponential function, for example, this leads to the same result as defining e^A directly through the power series of the exponential function.

The correspondence between classical and quantum mechanics consists in the requirement that the quantum mechanical operators $\hat{x}$, $\hat{p}$ corresponding to position x and momentum p satisfy the operator analogs of (2.1.12), that is,

$$[\hat{x}^i, \hat{x}^j] = 0 = [\hat{p}_i, \hat{p}_j] \quad \text{and} \quad [\hat{x}^j, \hat{p}_k] = i\hbar\delta^j_k, \qquad (2.1.82)$$

with the commutator of operators,

$$[A, B] = AB - BA \qquad (2.1.83)$$

in place of the Poisson bracket. The factor i in (2.1.82) comes from the fact that the commutator of two Hermitian operators is skew Hermitian.

$|x\rangle$ then denotes the state where the particle is localized at the point $x \in M$, i.e., the probability to find it at x is 1, and 0 elsewhere. When M is the real line $\mathbb{R}$, that is, one-dimensional, this is an eigenstate of the position operator $\hat{x}$, that is,

$$\hat{x}|x\rangle = |x\rangle x \qquad (2.1.84)$$

corresponding to the eigenvalue x. In $\mathbb{R}^d$, the components x^i, $i = 1, \ldots, d$, are the eigenvalues of the corresponding operators $\hat{x}^i$, and

$$\hat{x}^i|x\rangle = |x\rangle x^i. \qquad (2.1.85)$$

One should note that these operators are unbounded, and in our above L^2-space, the states $|x\rangle$ are represented by Dirac functionals $\delta(x)$, that is, they are *not* L^2-functions. Functional analysis provides concepts for making this entirely rigorous. In fact, according to spectral theory as presented above, we consider the Hilbert space $L^2(\mathbb{R})$ and write the operator as

$$A|\phi\rangle(x) = x|\phi\rangle(x), \qquad (2.1.86)$$

which then admits the spectral resolution

$$A = \int a\, dE(a) \qquad (2.1.87)$$

with

$$E(a)|\phi\rangle(x) = \begin{cases} |\phi\rangle(x) & \text{for } x \le a, \\ 0 & \text{for } x > a. \end{cases} \qquad (2.1.88)$$

Returning to (2.1.84), we see that the spectrum of the position operator $\hat{x}$ on $\mathbb{R}$ consists of the entire real line.

The established notation usually leaves out the hats, that is, writes x both for the position at a point in M and the corresponding operator on $\mathcal{H}$ that has been called $\hat{x}$ in (2.1.84). We shall also do that from this point on.

With these conventions, the Schrödinger equation (2.1.14) becomes

$$i\hbar \frac{\partial}{\partial t}|\psi(t)\rangle = H|\psi(t)\rangle. \qquad (2.1.89)$$

H, the Hamiltonian, here is a self-adjoint (Hermitian) operator, that is, an observable, in fact the most basic one of the whole theory.

The solution of (2.1.89) can be expressed by functional calculus as

$$|\psi(t)\rangle = e^{-\frac{i}{\hbar}tH}|\psi(0)\rangle. \qquad (2.1.90)$$

Here the exponential of $-H$ is defined through the usual power series of the exponential function, or better, via (2.1.81). Since H is Hermitian, the operators $e^{-\frac{i}{\hbar}tH}$ are unitary. Thus, the state $|\psi\rangle$ evolves by unitary transformations.

By taking $\frac{\partial}{\partial t}$ of (2.1.90), we see that formally it satisfies the Schrödinger equation (2.1.89), indeed. From (2.1.90), we also infer the semigroup property

$$|\psi(t)\rangle = e^{-\frac{i}{\hbar}(t-\tau)H}|\psi(\tau)\rangle = e^{-\frac{i}{\hbar}(t-\tau)H}|e^{-\frac{i}{\hbar}\tau H}|\psi(0)\rangle. \qquad (2.1.91)$$

Thus, the solution at time t is obtained from the solution at time τ by applying the solution operator for the remaining time $t - \tau$ ($0 < \tau < t$). We express the relation between (2.1.110) and (2.1.109) also by saying that $-\frac{i}{\hbar}H$ is the infinitesimal generator of the semigroup $e^{-\frac{i}{\hbar}tH}$. For an account of the mathematical theory of semigroups for partial differential equations, we refer to [63].

In this Schrödinger picture, the states evolve in time, whereas the observables don't. In the Heisenberg picture, this relation is reversed. The states are time-independent, whereas the operators representing the observables change in time, according to

$$i\hbar\frac{dA}{dt} = [A, H] \tag{2.1.92}$$

whose solution is

$$A(t) = e^{\frac{i}{\hbar}Ht}A(0)e^{-\frac{i}{\hbar}Ht}. \tag{2.1.93}$$

The Schrödinger and the Heisenberg picture are equivalent insofar as they yield the same probability density for the outcome of observations. This is expressed by the relation

$$\langle\phi(t)|A(0)|\psi(t)\rangle = \langle\phi(0)|e^{\frac{i}{\hbar}Ht}A(0)e^{-\frac{i}{\hbar}Ht}|\psi(0)\rangle = \langle\phi(0)|A(t)|\psi(0)\rangle \tag{2.1.94}$$

where we have used (2.1.90) and (2.1.93).

From (2.1.92), we also see that A is conserved precisely if it commutes with H. In that case, the quantity in (2.1.94) is constant in time, equaling

$$\langle\phi(0)|A(0)|\psi(0)\rangle. \tag{2.1.95}$$

Experimental interactions are formally described by the S-matrix. It is assumed that a state is prepared to have a definite particle content α for $t \to -\infty$ (that is, before the interaction takes place); this is the in state ψ_α^+. The interactions take place at finite time, and one then measures the out state ψ_β^- with particle content β for $t \to \infty$ (that is, after the interaction has taken place). Then, the probability amplitude for the transition is

$$S_{\beta\alpha} = \langle\psi_\beta^-, \psi_\alpha^+\rangle. \tag{2.1.96}$$

The $S_{\beta\alpha}$ are the components of the S-matrix. It is assumed here that these values are computed for complete sets of orthonormal in and out states, so that the S-matrix has to be unitary.

We consider once more the Hamiltonian (2.1.9). x is the position operator, and a particle that is located at a point $x \in M$ (note that we use the same symbol for the position and the position operator) is then in the eigenstate $|x\rangle$ of the position operator (note that this eigenstate in general will not be contained in the Hilbert space $L^2(M)$, but is instead given by a delta functional δ_x). If the particle is in the state $|x\rangle$ at time 0, then by the solution of the Schrödinger equation (2.1.90), at time t it will be in the state

$$e^{-\frac{i}{\hbar}tH}|x\rangle. \tag{2.1.97}$$

More generally, the probability amplitude $\langle x'', t''|x', t'\rangle$ that a particle starting at x' at time t' will be at x'' at the time $t'' > t'$ is given by

$$\langle x'', t''|x', t'\rangle = \langle x''|e^{-\frac{i}{\hbar}H(t''-t')}|x'\rangle, \tag{2.1.98}$$

the projection of the state $e^{-\frac{i}{\hbar}H(t''-t')}|x'\rangle$ obtained from the solution of the Schrödinger equation onto the state $|x''\rangle$.

By formal functional calculus, this is expressed as a Feynman functional integral

$$\langle x''|e^{-\frac{i}{\hbar}H(t''-t')}|x'\rangle = \int Dx \exp\left(\frac{i}{\hbar}L(x)\right) \tag{2.1.99}$$

$$= \int Dx \exp\left(\frac{i}{\hbar}\int_{\tau=t'}^{t''}\left(\frac{m}{2}|\dot{x}(\tau)|^2 - V(x(\tau))\right)d\tau\right) \tag{2.1.100}$$

for our standard example (2.1.7). Here, Dx symbolizes a formal measure on the space of all paths starting at time t' at x' and ending at time t'' at x'', according to the interpretation given in many texts. One should point out here, however, that this measure by itself is not well defined. What one can hope to attach a mathematical meaning to is only the entire integrand $Dx \exp(iL(x))$ in (2.1.99) as a functional measure on the path space. This is in analogy with the Wiener measure where one considers the probability density $p(x'', t''|x', t')$ for a particle starting at time t' at x' and ending up at time t'' at x'' under the influence of the potential $V(x)$, that is, governed by the Lagrangian (2.1.7),

$$F = \frac{m}{2}|\dot{x}|^2 - V(x). \tag{2.1.101}$$

The probability density evolves according to the heat equation

$$\frac{\partial \phi(x,t)}{\partial t} = m\Delta\phi(x,t) - V(x)\phi(x,t). \tag{2.1.102}$$

For comparison, we recall the Schrödinger equation (2.1.14) for the Lagrangian (2.1.7),

$$i\hbar\frac{\partial \phi(x,\tau)}{\partial \tau} = -\frac{\hbar^2}{2m}\Delta\phi(x,\tau) + V(x)\phi(x,\tau). \tag{2.1.103}$$

Let us assume $\hbar = m = 1$ to make the comparison a little simpler. Then, in fact, setting $\tau = -it$ transforms (2.1.103) into (2.1.102). Thus, the Schrödinger equation is the heat equation for imaginary time. With this change of time (called analytic continuation or Wick rotation in the physics literature), the corresponding functional integrals are also transformed into each other. This is useful at the formal level, but perhaps not as much so for the more detailed mathematical analysis.

Returning to (2.1.102), Wiener then showed that, under appropriate conditions on V, the solution can be represented as a path integral

$$p(x'', t''|x', t') = \int Dx \exp(-L(x))$$

$$= \int Dx \exp\left(-\int_{\tau=t'}^{t''}\left(\frac{m}{2}|\dot{x}(\tau)|^2 - V(x(\tau))\right)d\tau\right). \tag{2.1.104}$$

For the discussion to follow, it will be convenient to rewrite (2.1.104) slightly as

$$p(x'', t''|x', t') = \int [Dx]_{x',t'}^{x'',t''} \exp(-L(x)) \qquad (2.1.105)$$

to indicate the initial and terminal points of the paths over which we integrate. One then has the property

$$p(x'', t''|x', t') = \int_x p(x'', t''|x, t) p(x, t|x', t') dx \qquad (2.1.106)$$

for every $t' < t < t''$, which simply expresses the fact that every path leading from x' at time t' to x'' at time t'' has to pass through some x at the intermediate time t. Thus, we may cut the path at time t and integrate over all possible cutting points x. In terms of functional integrals, this becomes

$$p(x'', t''|x', t') = \int_x dx \int [Dx]_{x',t'}^{x,t} \exp(-L(x)) \int [Dx]_{x,t}^{x'',t''} \exp(-L(x)).(2.1.107)$$

The difference between (2.1.102) and (2.1.104) is the i versus the -1 in the exponent in the integral. In the Wiener case, the minus sign leads to a rapid dampening of the influence of those paths with large values of the Lagrangian action, and to a concentration of the functional measure near the minimum of the action. In the Feynman case, in contrast, paths with large values of the action cause rapid fluctuations in the integral, making the analysis substantially harder, see [2]. We do not enter the details here. For more on this, see [49].

In analogy to (2.1.106), we have the cutting relation

$$\langle x''|e^{-\frac{i}{\hbar}H(t''-t')}|x'\rangle$$
$$= \int [Dx]_{x',t'}^{x'',t''} \exp\left(\frac{i}{\hbar}L(x)\right)$$
$$= \int_x dx \int [Dx]_{x',t'}^{x,t} \exp\left(\frac{i}{\hbar}L(x)\right) \int [Dx]_{x,t}^{x'',t''} \exp\left(\frac{i}{\hbar}L(x)\right)(2.1.108)$$

for $t' < t < t''$.

Written more abstractly, this is the analog of (2.1.106),

$$\langle x'', t''|x', t'\rangle = \int_x dx \langle x'', t''|x, t\rangle\langle x, t|x', t'\rangle. \qquad (2.1.109)$$

This fits together well with the operator formalism as in (2.1.70). In particular, we can now insert a position operator (writing $x(t)$ in place of $\hat{x}(t)$ as announced above) and compute

$$\langle x'', t''|x(t)|x', t'\rangle = \int [Dx]_{x',t'}^{x'',t''} x(t) \exp\left(\frac{i}{\hbar}L(x)\right)$$
$$= \int_x dx \langle x'', t''|x, t\rangle x\langle x, t|x', t'\rangle. \qquad (2.1.110)$$

Here, the $x(t)$ in the integral is a number,[3] not an operator. That number in the integral then translates into the operator $x(t)$ in the inner product on the l.h.s. In physics, these inner products are viewed as the matrix elements of an infinite-dimensional matrix.

This process of cutting the path in the integral can be iterated, and we can insert two intermediate positions $x(t_1), x(t_2)$ (or more, but the principle emerges for two already), to get

$$\int [Dx]_{x',t'}^{x'',t''} x(t_1)x(t_2) \exp\left(\frac{i}{\hbar} L(x)\right) = \langle x'', t''|x(t_2)x(t_1)|x', t'\rangle. \qquad (2.1.111)$$

Here, in the integral, the temporal order of t_1 and t_2 is irrelevant because in the integral, $x(t_1), x(t_2)$ are real numbers.[4] In the r.h.s. of (2.1.111), however, they are operators, and since operators in general do not commute, the order does matter here. Since the paths are traversed in increasing time, we need to put them into the temporal order, that is, always apply the operator corresponding to the smaller time first. This is called temporal ordering. Formally, one can define the temporally ordered operator

$$T[x(t_2)x(t_1)] := \begin{cases} x(t_2)x(t_1) & \text{if } t_1 < t_2, \\ x(t_1)x(t_2) & \text{if } t_1 > t_2 \end{cases} \qquad (2.1.112)$$

and write the r.h.s. of (2.1.111) as

$$\langle x'', t''|T[x(t_2)x(t_1)]|x', t'\rangle. \qquad (2.1.113)$$

We may also write

$$T[x(t_2)x(t_1)] = \theta(t_2 - t_1)x(t_2)x(t_1) + \theta(t_1 - t_2)x(t_1)x(t_2) \qquad (2.1.114)$$

where

$$\theta(s) := \begin{cases} 1 & \text{for } s \geq 0, \\ 0 & \text{for } s < 0 \end{cases} \qquad (2.1.115)$$

is the Heaviside function. Considered as a functional, its derivative is the Dirac functional,

$$\frac{d}{ds}\theta(s) = \delta(s). \qquad (2.1.116)$$

---

[3]More precisely, when the path x takes its values in Euclidean space $\mathbb{R}^d$, $x(t)$ is a vector with d components. The operations in (2.1.110) and subsequent formulae are to be understood for each component. In particular, when we later on, in (2.1.111) and subsequently, insert expressions like $x(t_1)\cdots x(t_m)$, this is understood as the vector $(x^1(t_1)\cdots x^1(t_m), \ldots, x^d(t_1)\cdots x^d(t_m))$ obtained by componentwise multiplication.

[4]See the preceding footnote.

We now make some general observations about functional integrals of the form

$$\int [Dx]_{x',t'}^{x'',t''} \exp\left(\frac{i}{\hbar}L(x)\right). \tag{2.1.117}$$

In place of the position operator $x(t)$, we can also insert other operators $f(x)$ in (2.1.110). The formula becomes

$$\langle x'', t'' | f(x) | x', t' \rangle = \int [Dx]_{x',t'}^{x'',t''} f(x) \exp\left(\frac{i}{\hbar}L(x)\right). \tag{2.1.118}$$

Again, on the l.h.s., $f(x)$ stands for an operator, on the r.h.s. for a number.

The analogy between ordinary (Gaussian) integrals and functional integrals says that the finitely many ordinary degrees of freedom, the coordinate values of the integration variable, are replaced by the infinitely many function values $x(t)$. Therefore, integration by parts should yield that

$$0 = \int [Dx]_{x',t'}^{x'',t''} \frac{\delta}{\delta x(t)} \exp\left(\frac{i}{\hbar}L(x)\right), \tag{2.1.119}$$

that is, the integral of a total derivative vanishes. This yields

$$0 = \frac{i}{\hbar} \int [Dx]_{x',t'}^{x'',t''} \exp\left(\frac{i}{\hbar}L(x)\right) \frac{\delta}{\delta x(t)} L(x). \tag{2.1.120}$$

Recalling (2.1.118), this is written as

$$\langle x'', t'' | \frac{\delta L(x)}{\delta x(t)} | x', t' \rangle = 0. \tag{2.1.121}$$

Now, $\frac{\delta L(x)}{\delta x(t)}$ represents the Euler–Lagrange operator (see also (2.3.8) below), and

$$\frac{\delta L(x)}{\delta x(t)} = 0 \tag{2.1.122}$$

is the classical equation of motion. Comparing (2.1.122) and (2.1.121), we see that the classical equation of motion is translated into an operator equation in the quantum mechanical picture. In this interpretation, x', t' and x'', t'' represent arbitrary initial and final conditions for our paths $x(t)$.

Returning to our integration by parts, (2.1.119) generalizes to

$$\frac{i}{\hbar} \int [Dx]_{x',t'}^{x'',t''} \exp\left(\frac{i}{\hbar}L(x)\right) \frac{\delta L(x)}{\delta x(t)} f(x)$$

$$= \int [Dx]_{x',t'}^{x'',t''} \frac{\delta}{\delta x(t)} \left(\exp\left(\frac{i}{\hbar}L(x)\right)\right) f(x)$$

$$= -\int [Dx]_{x',t'}^{x'',t''} \exp\left(\frac{i}{\hbar}L(x)\right) \frac{\delta}{\delta x(t)} f(x). \tag{2.1.123}$$

When none of the fields (see below) contained in $f(x)$ is evaluated at time t, the r.h.s. vanishes. This, of course, confirms the interpretation of (2.1.122) as a quantum mechanical operator equation.

Naturally, we are now curious what happens when some of the fields are present at time t. So, we insert $x(t_0)$ for some $t' < t_0 < t''$. This yields

$$\frac{i}{\hbar} \int [Dx]_{x',t'}^{x'',t''} x(t_0) \exp\left(\frac{i}{\hbar} L(x)\right) \frac{\delta}{\delta x(t)} L(x)$$

$$= \int [Dx]_{x',t'}^{x'',t''} \frac{\delta}{\delta x(t)} \left(\exp\left(\frac{i}{\hbar} L(x)\right)\right) x(t_0)$$

$$= -\int [Dx]_{x',t'}^{x'',t''} \exp\left(\frac{i}{\hbar} L(x)\right) \frac{\delta}{\delta x(t)} x(t_0)$$

$$= -\int [Dx]_{x',t'}^{x'',t''} \exp\left(\frac{i}{\hbar} L(x)\right) \delta(t - t_0). \tag{2.1.124}$$

As an operator equation, this is interpreted as

$$\frac{i}{\hbar} x(t_0) \frac{\delta}{\delta x(t)} L(x) = \delta(t - t_0). \tag{2.1.125}$$

As remarked above, $\frac{\delta L(x)}{\delta x(t)}$ represents the Euler–Lagrange operator (2.3.8). We consider here the example $\frac{d^2}{dt^2}$, and the classical Euler–Lagrange equation becomes the operator equation

$$\frac{d^2}{dt^2} x(t) = 0. \tag{2.1.126}$$

From (2.1.114), (2.1.116), (2.1.126), (2.1.45), we obtain

$$\frac{d^2}{dt^2} T[x(t_2)x(t_1)] = \delta(t_2 - t_1) \left[\frac{d}{dt} x(t_1), x(t_1)\right] = -i\hbar\delta(t_2 - t_1) \tag{2.1.127}$$

using the Heisenberg commutation relation (2.1.12) for the last equation. This, of course, coincides with (2.1.125). In Sect. 2.5.1 below, this equation will lead us to the normal ordering scheme for operators.

A recent reference on path integrals is [112].

2.1.4 Quasiclassical Limits

In this section, we briefly discuss some analytical aspects of the relationship between classical and quantum mechanics.

In classical mechanics, stable equilibria are characterized by the principle of locally minimal potential energy, whereas dynamical processes are described by the principle of stationary motion. Both are variational principles. We consider a physical system with d degrees of freedom $x^1, \ldots, x^d$. We want to determine the motion of the system by expressing the x^i as functions of the time t. The mechanical properties of the system are described by the kinetic and the potential energy. The kinetic energy is typically of the form

$$T = \sum_{i,j=1}^{d} A_{ij}(x^1, \ldots, x^d, t) \dot{x}^i \dot{x}^j. \tag{2.1.128}$$

Thus, T is a function of the velocities $\dot{x}^1, \ldots, \dot{x}^d$, the coordinates $x^1, \ldots, x^d$, and time t; often, T does not depend explicitly on t, and one may then investigate equilibria. Here, T is a quadratic form in the generalized velocities $\dot{x}^1, \ldots, \dot{x}^d$.

The potential energy is of the form

$$V = V(x^1, \ldots, x^d, t), \tag{2.1.129}$$

that is, it does not depend on the velocities.

In order not to have to worry about the justification of taking various derivatives, we assume that V and T are of class C^2.

Hamilton's principle now postulates that motion between two points in time t_0 and t_1 occurs in such a way that the Lagrangian action

$$L(x) := \int_{t_0}^{t_1} (T - V) dt \tag{2.1.130}$$

is stationary in the class of all functions $x(t) = (x^1(t), \ldots, x^d(t))$ with fixed initial and final states $x(t_0)$ and $x(t_1)$ respectively. The Lagrangian action is the integral over the Lagrangian

$$F(t, x, \dot{x}) := T - V, \tag{2.1.131}$$

the difference between kinetic and potential energy.

Thus, one does not necessarily look for a minimum under all motions which carry the system from an initial state to a final state, but only for a stationary value of the integral. For such a stationary motion, the Euler–Lagrange equations hold:

$$\frac{d}{dt} \frac{\partial T}{\partial \dot{x}^i} - \frac{\partial}{\partial x^i}(T - V) = 0 \quad \text{for } i = 1, \ldots, d. \tag{2.1.132}$$

If V and T do not explicitly depend on the time t, then equilibrium states are constant in time, that is, $\dot{x}^i = 0$ for $i = 1, \ldots, d$, and hence $T = 0$, therefore by (2.1.132)

$$\frac{\partial V}{\partial x^i} = 0 \quad \text{for } i = 1, \ldots, d. \tag{2.1.133}$$

Thus, in a state of equilibrium, V must have a critical point, and in order for this equilibrium to be stable, V must even have a minimum there. That minimum, however, need not be unique. For example, when the state space is simply the real line

$\mathbb{R}$, and

$$V(x) = (x^2 - a^2)^2 \tag{2.1.134}$$

for some $a \in \mathbb{R}$, the classical equilibrium states are $x = \pm a$.

Quantum mechanically, we have an L^2-function $\phi : \mathbb{R} \to \mathbb{C}$ (but, for simplicity, we shall consider real-valued functions ϕ in the present section) with the normalization

$$\|\phi\|_{L^2} = 1 \tag{2.1.135}$$

and the asymptotic behavior

$$\lim_{x \to \pm \infty} \phi(x) = 0. \tag{2.1.136}$$

The potential energy $V(x)$ is now replaced by the energy

$$\int_{\mathbb{R}} \left(\frac{\hbar^2}{2} \left| \frac{d\phi(x)}{dx} \right|^2 + V(x)|\phi(x)|^2 \right) dx. \tag{2.1.137}$$

The corresponding Euler–Lagrange equation is

$$\left(-\frac{\hbar^2}{2} \frac{d^2}{dx^2} + V(x) \right) \phi = 0. \tag{2.1.138}$$

Since there is no kinetic term in (2.1.137), quantum mechanics tries to find the eigenfunctions of the operator in (2.1.138), the Hamiltonian. That is, we look for solutions of

$$\left(-\frac{\hbar^2}{2} \frac{d^2}{dx^2} + V(x) \right) \phi_i = E_i \phi_i. \tag{2.1.139}$$

The eigenvalues E_i are the energy levels. A solution ϕ_0 for the smallest possible energy E_0 corresponds to the vacuum. E_0 is positive since the potential V is positive, see (2.1.134). The solution ϕ_0 (normalized by $\|\phi_0\|_{L^2} = 1$ according to (2.1.135)) is symmetric, that is, $\phi_0(x) = \phi_0(-x)$, with maxima at $\pm a$, a local minimum at 0, and asymptotic decay $\lim_{x \to \pm \infty} \phi(x) = 0$ required by (2.1.136). The eigenfunctions for different eigenvalues are L^2-orthogonal. In particular, the eigenfunction ϕ_1 for the second smallest energy level E_1 satisfies $\int \phi_0 \phi_1 dx = 0$. It is antisymmetric, that is $\phi_1(x) = -\phi_1(-x)$ and thus changes sign at $x = 0$. In the quasiclassical limit, that is, for $\hbar \to 0$, we have

$$E_0, E_1 \sim a\hbar \tag{2.1.140}$$

and

$$E_1 - E_0 \sim c_0 \exp\left(-\frac{c_1}{\hbar} \right) \tag{2.1.141}$$

for constants a, c_0, c_1. Thus, the difference between these energy levels goes to 0 exponentially. Therefore, in that quasiclassical limit, the energy levels of ϕ_0 and ϕ_1

become indistinguishable. Thus, for $\hbar \to 0$, the limits of $\phi_\pm := \frac{1}{\sqrt{2}}(\phi_0 \pm \phi_1)$ also become minima, with

$$\lim_{\hbar \to 0} |\phi_\pm|^2 = \delta(x \mp a). \qquad (2.1.142)$$

ϕ_+ and ϕ_- break the symmetry between a and $-a$. Classically, any linear combination of $\delta(x - a)$ and $\delta(x + a)$ is a possible minimum. The quantum mechanical vacuum, however, is symmetric.

It is also instructive to consider a nonlinear problem. We take

$$\int \left(\frac{\hbar^2}{2} \left| \frac{d\phi(x)}{dx} \right|^2 + W(\phi(x)) \right) dx. \qquad (2.1.143)$$

This time, ϕ need not be real-valued, but could assume values in some other space, like a Riemannian manifold N. We first consider the real-valued case. The domain, denoted by M, however, is allowed to be of higher dimension. We suppose again that the potential W has two minima (W is then called a two-well potential). Now, in the quasiclassical limit, we do not obtain a concentration at two points, the two minima, in the domain, but rather the concentration at two values of ϕ. This time, in contrast to the linear case, the symmetry can also be broken for $\hbar > 0$. Since this is not a linear problem, we no longer have the concept of eigenfunctions available. For $\hbar \to 0$, the solution becomes piecewise constant, the values being the two minima of W, of course. When one imposes suitable constraints, by a result of Modica [82], the set of discontinuity of the limit for $\hbar \to 0$ of the solutions for $\hbar > 0$ is a hypersurface of constant mean curvature in the domain. Of course, this is meaningful only for a higher-dimensional domain.

When the domain is one-dimensional, that is, the real line $\mathbb{R}$, but the target is of higher dimension, we may have quantum mechanical tunneling solutions, i.e.,

$$\lim_{x \to \pm\infty} \phi(x) = a_\pm \qquad (2.1.144)$$

between the vacua, that is, minima of W, denoted by a_+, a_-. These tunneling solutions are gradient flow lines of W when the target is a Riemannian manifold.

There exist some generalizations of this problem that lead to analytical constructions of great interest:

1. Let L be a real line bundle over $M \backslash S$ where S is a submanifold of codimension 2 in the domain M, with prescribed holonomy for L around S. A section of L then has a zero set of codimension 1 in M with boundary S. Considering the quasiclassical limit of those zero sets for minimizers of the above functional for sections of L yields a minimal hypersurface in M with boundary S, according to [44].
2. Let L now be a complex line bundle over M. The above functional then leads to a vortex equation of the type studied by Taubes [99], Bethuel et al. [13], and Ding et al. [28–30].

2.2 Lagrangians

2.2.1 Lagrangian Densities for Scalars, Spinors and Vectors

A type of particle is represented by a vector bundle E over some Lorentz manifold M. The particle transforms according to some representation of the Lorentz group or its double cover, the spin group.[5] Thus, it transforms as a tensor or as a spinor. The states of collections of such particles are represented by sections ψ of E, so-called fields.[6]

We are considering here the semiclassical situation, i.e., before field quantization, and so ψ has to satisfy the Euler–Lagrange equations of some action functional that is invariant under the representation of the Lorentz or spin group according to which the particle transforms. In addition, there are internal symmetries that affect only the values of the fields, but not of the coordinates, and leave the action invariant. In fact, the symmetries and certain general considerations often suffice to allow us to construct the appropriate Lagrangian for the action as we shall see.

Notation: $\partial_\mu = \frac{\partial}{\partial x^\mu}$, $\partial^\mu = g^{\mu\nu}\partial_\nu$.

For the moment, we can think of $g_{\mu\nu}$ as a (Lorentz) metric on $\mathbb{R}^{1,3}$, and the indices μ, ν then run from 0 to 3.

We consider the action functional (Lagrangian)[7]

$$S(\phi) = \int \left\{ \frac{1}{2}\partial_\mu\phi\partial^\mu\phi - \frac{1}{2}m^2\phi^2 \right\} \sqrt{-g(x)}d(x) \tag{2.2.1}$$

for a free scalar field, with the Lagrangian density

$$F(\phi, D\phi) = \frac{1}{2}\partial_\mu\phi\partial^\mu\phi - \frac{1}{2}m^2\phi^2, \tag{2.2.2}$$

on $\mathbb{R}^{1,3}$, or, more generally, on some Riemannian or Lorentzian manifold. (Note that in (2.2.1), we use $\sqrt{-g(x)}d(x)$ for the volume form, since we are assuming a Lorentzian metric. Subsequently, when we switch to the Riemannian case, the minus sign has to be deleted.)

The corresponding Euler–Lagrange equation is the Klein–Gordon equation

$$\Box\phi + m^2\phi = 0$$

where $\Box$ is the Minkowski Laplacian (1.1.106).

---

[5]In fact, according to Wigner's principle as explained in Sect. 1.3.4, we should consider a particle as an irreducible unitary representation not only of the Lorentz or spin group, but of the Poincaré group or the double covering $Sl(2, \mathbb{C}) \ltimes \mathbb{R}^{1,3}$. While this is fundamental for determining the types of possible elementary particles from the theory of group representations, in this section, we shall be mainly concerned with internal symmetries that arise from invariance w.r.t. to the action of some compact group.

[6]A section represents a state containing possibly several particles of a given type, since in quantum field theory, particle numbers need not be preserved.

[7]In the mathematical literature, an action functional is often called a Lagrange functional.

Remark

1. As a classical action functional for a field, we consider the action for a particle $q(t) = (q^1(t), \ldots, q^m(t))$ with m degrees of freedom

$$\int \sum_{j=1}^{m} \left\{ \frac{1}{2} (\dot{q}^j(t)\dot{q}_j(t) - m^2 q^j(t)q_j(t)) \right\} dt. \tag{2.2.3}$$

When we compare this with our quantum field theoretic setting, we see that the index j corresponds to the spatial variable (x^1, x^2, x^3) above. In this sense, $\phi(t, x)$ is a particle with infinitely many degrees of freedom, one degree of freedom for each point of M.

2. In the physics literature, the field ϕ in (2.2.1), (2.2.2) is usually taken as complex valued instead of real valued, that is, one considers

$$S_1(\phi) = \int \left\{ \frac{1}{2} \partial_\mu \phi \partial^\mu \bar{\phi} - \frac{1}{2} m^2 |\phi|^2 \right\} \sqrt{-g(x)} d(x), \tag{2.2.4}$$

in line with the basic formalism of quantum mechanics, see (2.1.13). Our reason for starting with a real valued ϕ here is, besides its simplicity, that this is better suited for subsequent generalizations to nonlinear models where the field will take its values in a Riemannian manifold.

We now turn to Lagrangians for spinors.
For two left-handed spinors (see (1.3.49)) ϕ, χ,

$$\phi\chi := \varepsilon_{\alpha\beta}\phi^\alpha \chi^\beta$$

transforms as a scalar under the spinor representation, see (1.3.56).
Similarly

$$\phi^\alpha \sigma^\mu_{\alpha\dot\alpha} \bar\chi^{\dot\alpha}$$

transforms as a vector, for $\mu = 0, 1, 2, 3$, see (1.3.57). We may then write a Lagrangian for a left-handed spinor ϕ as

$$F = \text{Re}(i\phi\sigma^\mu \partial_\mu \bar\phi + 2m\phi\phi)$$

$$= \frac{i}{2}(\phi\sigma^\mu \partial_\mu \bar\phi - \partial_\mu \phi \sigma^\mu \bar\phi) + m(\phi\phi + \bar\phi\bar\phi) \tag{2.2.5}$$

(here, $\bar\phi$ is the complex conjugate of ϕ—subsequently, we shall employ a somewhat different convention when we consider full spinors).
The equation of motion for

$$S(\phi) = \int F(\phi) \tag{2.2.6}$$

is

$$i\partial_\mu \phi \sigma^\mu - m\bar{\phi} = 0, \tag{2.2.7}$$

or equivalently

$$i\sigma^\mu \partial_\mu \bar{\phi} + m\phi = 0. \tag{2.2.8}$$

In quantum field theory (QFT), charged particles correspond to complex-valued fields, and the Lagrangian has to remain invariant under multiplication of the fields by $e^{i\lambda}(\lambda \in \mathbb{R})$ since the phase is not observable. Instead of imposing the normalization $\|\phi\| = 1$, we can then consider states as corresponding to lines in a Hilbert space.

The preceding Lagrangian satisfies this invariance property only for $m = 0$. Since it does not have this property in general, it corresponds to a neutral fermion. In the standard model to be discussed below, these neutral fermions are the neutrinos.

In order to obtain a Lagrangian for charged fermions, we need full spinors

$$\psi = \begin{pmatrix} \psi_L \\ \psi_R \end{pmatrix}.$$

Then in the Weyl representation,

$$\bar{\psi}\gamma^\mu \psi$$

transforms as a vector, see (1.3.61).

The Dirac–Lagrangian is then

$$F = i\bar{\psi}\gamma^\mu \partial_\mu \psi - m\bar{\psi}\psi = i\langle \bar{\psi}, \cancel{D}\psi \rangle - m\bar{\psi}\psi \tag{2.2.9}$$

(recalling the Dirac operator $\cancel{D}$ defined in (1.3.22)). The mass term mixes the left and the right spinor, since $\bar{\psi}\psi = \bar{\psi}_L\psi_R + \bar{\psi}_R\psi_L$. This time, we do have invariance under multiplication of ψ by $e^{i\lambda}$ for constant real λ also in the general case $m \neq 0$.

The corresponding Dirac equation is

$$i\gamma^\mu \partial_\mu \psi - m\psi = 0. \tag{2.2.10}$$

Perhaps the factor i in (2.2.9) in front of the Dirac operator $\cancel{D} = \gamma^\mu \partial_\mu$ needs some explanation. The reason is that, upon integration, the corresponding term is purely imaginary, and the factor i then makes it real. It is instructive to consider an example, and since we shall mainly investigate the Riemannian in place of the Lorentzian setting in the sequel, we shall also use a Riemannian example. As in our treatment of the supersymmetric sigma model below (2.4.3), we consider the two-dimensional case and use the following representation of the Clifford algebra $Cl(2, 0)$:

$$e_1 \to \begin{pmatrix} -1 & 0 \\ 0 & 1 \end{pmatrix}, \qquad e_2 \to \begin{pmatrix} 0 & -1 \\ -1 & 0 \end{pmatrix}, \tag{2.2.11}$$

which is different from the one described in Sect. 1.3.2 (but of course equivalent to it). A spinor ω is thus identified with an element $(\omega_1 = \alpha_1 + i\beta_1, \omega_2 = \alpha_2 + i\beta_2)$ of

$\mathbb{C}^2$. In local coordinates x, y, then

$$\gamma^\mu \partial_\mu \omega = \begin{pmatrix} -1 & 0 \\ 0 & 1 \end{pmatrix} \frac{\partial}{\partial x} \begin{pmatrix} \omega_1 \\ \omega_2 \end{pmatrix} + \begin{pmatrix} 0 & -1 \\ -1 & 0 \end{pmatrix} \frac{\partial}{\partial y} \begin{pmatrix} \omega_1 \\ \omega_2 \end{pmatrix}$$

$$= \begin{pmatrix} -\frac{\partial \alpha_1}{\partial x} + i \frac{\partial \beta_1}{\partial x} + \frac{\partial \alpha_2}{\partial y} + i \frac{\partial \beta_2}{\partial y} \\ \frac{\partial \alpha_2}{\partial x} + i \frac{\partial \beta_2}{\partial x} - \frac{\partial \alpha_1}{\partial y} - i \frac{\partial \beta_1}{\partial y} \end{pmatrix} \tag{2.2.12}$$

and

$$\bar{\omega}\gamma^\mu \partial_\mu \omega = -\alpha_1 \frac{\partial \alpha_1}{\partial x} - \alpha_1 \frac{\partial \alpha_2}{\partial y} - \beta_1 \frac{\partial \beta_1}{\partial x} - \beta_1 \frac{\partial \beta_2}{\partial y} + \alpha_2 \frac{\partial \alpha_2}{\partial x} - \alpha_2 \frac{\partial \alpha_1}{\partial y}$$

$$+ \beta_2 \frac{\partial \beta_2}{\partial x} - \beta_2 \frac{\partial \beta_1}{\partial y} + i \left(\beta_1 \frac{\partial \alpha_1}{\partial x} - \alpha_1 \frac{\partial \alpha_2}{\partial y} - \alpha_1 \frac{\partial \beta_1}{\partial x} + \beta_1 \frac{\partial \alpha_2}{\partial y} - \beta_2 \frac{\partial \alpha_2}{\partial x} \right.$$

$$\left. + \beta_2 \frac{\partial \alpha_1}{\partial y} + \alpha_2 \frac{\partial \beta_2}{\partial x} - \alpha_2 \frac{\partial \beta_1}{\partial y} \right). \tag{2.2.13}$$

Upon integration, the real part vanishes by integration by parts, and only the imaginary part remains. This comes about because the coefficients of ω commute. Were they to anticommute, only the real part would remain. We are making this observation here because in our subsequent treatment of supersymmetry, we shall use spinor fields with anticommuting coefficients.

We have now seen action functionals for scalars and spinors, where these names describe the transformation behavior under Lorentz transformations, i.e., coordinate changes. An electromagnetic field, however, is described by a potential that transforms as a vector or covector. We consider

$$A = A_\mu(x)dx^\mu.$$

A is called a vector particle, because A^μ transforms as a vector. Mathematically, A is a connection, see (1.2.12), on a vector bundle with fiber $\mathbb{C}$ and the Abelian structure group $U(1) = SO(2)$. We also recall the transformation behavior (1.2.32).

The field strength is described by the tensor

$$F_{\mu\nu} = \partial_\mu A_\nu - \partial_\nu A_\mu,$$

(2 times) the[8] curvature of the connection A, see (1.2.23) (note that the brackets $[A_\mu, A_\nu]$ vanish here, because the structure group is Abelian), and the corresponding Lagrangian is the Maxwell density (the Abelian case of the Yang–Mills density)

$$-\frac{1}{4} F_{\mu\nu} F^{\mu\nu} = \frac{1}{2}(\partial_\mu A_\nu \partial^\mu A^\nu - \partial_\mu A_\nu \partial^\nu A^\mu). \tag{2.2.14}$$

---

[8] Note the different conventions between the present section and Sect. 1.2.2.

An important property of this Lagrangian is the gauge invariance, namely its invariance under replacing A_μ by

$$A_\mu + \partial_\mu \xi$$

where ξ is a scalar function. This is the present, Abelian, version of (1.2.13), (1.2.32). Of course, the field strength $F_{\mu\nu}$ is already invariant under such a gauge transformation (see (1.2.25), (1.2.33) for the general result).

The equations of motion for

$$S(A) = -\frac{1}{4} \int F_{\mu\nu} F^{\mu\nu} \tag{2.2.15}$$

are

$$\partial_\mu F^{\mu\nu} = 0. \tag{2.2.16}$$

If we add a "mass term"

$$m^2 A_\mu A^\mu,$$

then the gauge invariance no longer holds.

As described, the mathematical interpretation of A is that of a covariant derivative for sections of a line bundle, see Sect. 1.2.2. Thus, for a scalar field ϕ taking values in this bundle, we put

$$(d_A \phi)_\mu = \partial_\mu \phi + A_\mu \phi,$$

and we may consider the interaction Lagrangian

$$\frac{1}{2}(\partial_\mu \phi + A_\mu \phi)(\partial^\mu \phi^* + (A^\mu \phi)^*) = \frac{1}{2}\|d_A \phi\|^2. \tag{2.2.17}$$

Here, we assume that the line bundle is Hermitian, and for simplicity, we write the metric as $\|\phi\|^2 = \phi\phi^*$; of course, in general this only holds in suitable coordinates; we also assume that A is unitary w.r.t. this metric—we shall return to this point in a moment.

The replacement of

$$\partial_\mu \phi \quad \text{with } \partial_\mu \phi + A_\mu \phi \tag{2.2.18}$$

is for the following reason. The Lagrangian

$$\frac{1}{2}\partial_\mu \phi \partial^\mu \phi^* - \frac{1}{2}m^2 \phi\phi^*$$

is invariant under $U(1)$, i.e., under replacements

$$\phi \mapsto e^{i\vartheta}\phi \quad \text{with } \vartheta \in \mathbb{R}.$$

It thus has a global internal symmetry. It is not invariant, however, under general local symmetries, i.e.,

$$\phi \mapsto e^{i\vartheta(x)}\phi$$

if $\vartheta(x)$ is a nontrivial function of x. However, if, according to (1.2.14), we also replace[9]

$$A_\mu \mapsto A_\mu - i\partial_\mu\vartheta,$$

then the above interaction Lagrangian (2.2.17) remains invariant. Thus, we have a gauge invariant Lagrangian. The procedure (2.2.18) of replacing an ordinary by a covariant derivative is called minimal coupling.

In fact, we have a free parameter here: We consider the exterior derivative d as the trivial connection ("vacuum") on the trivial bundle $M \times \mathbb{R}$. We can then view the affine space of connections A as the vector space $\Omega^1(M)$ of 1-forms. We can therefore multiply A by some factor q and choose the covariant derivative

$$D_A := \partial + qA \qquad (2.2.19)$$

and gauge transform A to

$$A - \frac{i}{q}\partial\vartheta. \qquad (2.2.20)$$

q here is interpreted as the charge of the electromagnetic field. It is the Noether charge associated to the $U(1)$ gauge symmetry.

The full Lagrangian for a complex scalar field ϕ interacting with an electromagnetic field A is

$$\frac{1}{2}\|d_A\phi\|^2 - \frac{1}{2}m^2\|\phi\|^2 + \frac{1}{4q^2}\|F\|^2. \qquad (2.2.21)$$

The same discussion applies to spinor fields ψ, and we may form the interaction Lagrangian

$$i\bar\psi\gamma^\mu(\partial_\mu + A_\mu)\psi - m\bar\psi\psi + \frac{1}{4q^2}\|F\|^2. \qquad (2.2.22)$$

Let us see the details once more: Replacing $\psi(x)$ by $e^{i\vartheta(x)}\psi(x)$ changes the spinor Lagrangian

$$i\bar\psi\gamma^\mu\partial_\mu\psi - m\bar\psi\psi \qquad (2.2.23)$$

by $-\bar\psi\gamma^\mu\psi\partial_\mu\vartheta$, and this is again compensated when we replace ∂_μ by $\partial_\mu + qA_\mu$ and require that A transforms to $A - \frac{i}{q}\partial\vartheta$ as before. Thus,

$$i\bar\psi\gamma^\mu(\partial_\mu + A_\mu)\psi - m\bar\psi\psi \qquad (2.2.24)$$

remains gauge invariant.

---

[9]Note that the convention here is different from the one in Sect. 1.2.3; here, elements of the Lie algebra $\mathfrak{u}(1)$ of $U(1)$, and similarly of other Lie algebras $\mathfrak{g}$, are written as $i\vartheta$, with a *real* ϑ. This will lead to various factors i and -1 when compared to Sect. 1.2.3. This is the standard convention employed in the physics literature.

As was realized by Yang and Mills,[10] this can be generalized to an arbitrary internal symmetry group G with Lie algebra $\mathfrak{g}$, and a field ϕ that takes its value in a vector bundle (or, similarly, in a spinor bundle—the physically more important case, see Sect. 2.2.3 below)[11] with structure group G. The mathematical formalism for this has been described in Sect. 1.2.3. In abstract physical terms, the gauge principle says that the symmetries should determine the forces. The particles conveying these forces are called gauge bosons.

To implement this, we simply consider $A = A_\mu dx^\mu$, a 1-form with values in $\mathfrak{g}$, and form the covariant derivative (1.2.12) of the field ϕ, a section of the vector bundle with structure group G on which A operates as a covariant derivative,

$$d_A\phi = d\phi + A\phi.$$

The replacements

$$\phi(x) \mapsto g(x)\phi(x), \quad \text{with } g(x) \in G \text{ for all } x,$$

i.e., g is an element of the group of gauge transformations, and (1.2.32), that is,

$$A \mapsto gAg^{-1} - (\partial g)g^{-1}$$

then leave

$$\|d_A\phi\|^2$$

invariant (assuming of course that the metric $\|\cdot\|$ is G-invariant).

The gauge field strength is now (two times) the curvature (1.2.22), (1.2.23)

$$F_{\mu\nu} = \partial_\mu A_\nu - \partial_\nu A_\mu + [A_\mu, A_\nu]$$

where $[\cdot, \cdot]$ is the Lie algebra bracket of $\mathfrak{g}$.

As before, we may form the Lagrangian involving the Yang–Mills action (1.2.34) and coupling it with the action for the field ϕ

$$\frac{1}{2}\|d_A\phi\|^2 - \frac{1}{2}m^2\|\phi\|^2 - \frac{1}{4q^2} \text{Tr}\, F_{\mu\nu} F^{\mu\nu}. \tag{2.2.25}$$

The same discussion applies to spinor fields ψ with values in a vector bundle on which G acts, that is, sections of $S \otimes E$ for some vector bundle E over M with a G-action, and we may form the interaction Lagrangian

$$i\bar{\psi}\gamma^\mu(\partial_\mu + A_\mu)\psi - m\bar{\psi}\psi - \frac{1}{4q^2}\text{Tr}\, F_{\mu\nu} F^{\mu\nu}. \tag{2.2.26}$$

---

[10]Such ideas were first conceived by Hermann Weyl.

[11]Of course, a spinor bundle is a vector bundle, but in physics, it is important to distinguish between vector and spinor representations, that is, whether a representation of the spin group lifts to one of the orthogonal group or not.

If the representation of G on our vector bundle is not irreducible, but decomposes into subrepresentations indexed by j, we may use a more general Lagrangian for the gauge field strength because we can take a combination

$$-\sum_j \gamma_j \operatorname{Tr} F^{(j)}_{\mu\nu} F^{(j)\mu\nu}.$$

The γ_j are the so-called coupling constants. Mathematically, they parametrize the ad-invariant bilinear forms on the Lie algebra of G.

Important remark: It is undesirable to have too many constants whose values are not theoretically deduced, but can only be experimentally determined. As we have just seen, such a situation comes about if the representation of G under consideration is not irreducible. One possible solution of this problem would be to suppose that there is some larger group $\tilde{G} \supset G$ in the background with an irreducible representation that induces the (reducible) representation of G, and so determines all the constants except one. It may be possible that the symmetry group $\tilde{G}$ cannot be experimentally observed because of a symmetry-breaking mechanism that reduces $\tilde{G}$ to G. The Higgs mechanism, to be described below, is such a mechanism.

Let us recapitulate that A is a 1-form with values in $\mathfrak{g}$, and the symmetries therefore are two-fold: Under Lorentz or space–time symmetries, the 1-form part is transformed, whereas under local internal symmetries (i.e., those coming from G), the $\mathfrak{g}$ part is affected. Similarly, ϕ transforms as a scalar under space–time symmetries and by the action of G under local internal symmetries.

More generally, one may also introduce some nonlinearities into the ϕ and ψ equations by adding some polynomial terms to the Lagrangian. These polynomial terms, however, are constrained by the requirement of renormalizability. In dimension 4, the most general renormalizable Lagrangian is

$$-\frac{1}{4q^2} \operatorname{Tr} F_{\mu\nu} F^{\mu\nu} + \frac{1}{2}\|d_A\phi\|^2 - \frac{1}{2}m^2\|\phi\|^2 + i\bar{\psi}\gamma^\mu(\partial_\mu A_\mu)\psi$$

$$+ g_1\|\phi\|^4 + g_2\phi\bar{\psi}\psi + \text{lower-order terms.} \qquad (2.2.27)$$

Here, g_1, g_2, like q, are coupling constants. The term $\phi\bar{\psi}\psi$ is called a Yukawa term. A version of (2.2.27) is also the Lagrangian of the standard model to which we shall turn after a brief discussion of the scaling behavior of Lagrangians. The gauge group of the standard model is $SU(3) \times SU(2) \times U(1)$.

2.2.2 Scaling

An important criterion for field theories in physics and variational problems in mathematics is their scaling behavior. That means that one scales the independent variables

$$x \to \lambda^{-1}x =: y \qquad (2.2.28)$$

where x is n-dimensional, $x \in \mathbb{R}^n$, and $\lambda > 0$, and computes the resulting scaling behavior of the integrals of the fields. The starting point is the scaling of the volume form

$$d^n y = \lambda^{-n} dx. \tag{2.2.29}$$

Also,

$$\frac{\partial}{\partial y} = \lambda \frac{\partial}{\partial x}. \tag{2.2.30}$$

Putting $\phi_\lambda(y) := \phi(\lambda y) = \phi(x)$, one obtains

$$\int_{\mathbb{R}^n} |d\phi_\lambda(y)|^p d^n y = \lambda^{p-n} \int_{\mathbb{R}^n} |d\phi(x)|^p d^n x \tag{2.2.31}$$

and

$$\int_{\mathbb{R}^n} |\phi_\lambda(y)|^q d^n y = \lambda^q \int_{\mathbb{R}^n} |\phi(x)|^q d^n x \tag{2.2.32}$$

for exponents $p, q > 0$. Therefore, the L^q-norm of ϕ_λ, $(\int_{\mathbb{R}^n} |\phi_\lambda(y)|^q d^n y)^{1/q}$ has a scaling behavior dominated by the L^p-norm of the derivative $d\phi_\lambda$, $(\int_{\mathbb{R}^n} |d\phi_\lambda(y)|^p d^n y)^{1/p}$, if

$$q \le \frac{np}{n-p} \quad \text{for } p < n. \tag{2.2.33}$$

This is exploited in the Sobolev embedding theorem, see e.g. [63]. For instance, in the most important case $p = 2$, in dimension 2, any polynomial in ϕ is controlled by $\int |d\phi|^2$, in dimension 3, a polynomial of order ≤ 6, and in dimension 4, only those of order ≤ 4.

In the light of (2.2.29), (2.2.30), the scaling law (2.2.31) can also be interpreted in the way that $\int_{\mathbb{R}^n} |d\phi|^2$ remains invariant if the field ϕ is scaled as

$$\phi \to \lambda^{\frac{n-2}{2}} \phi. \tag{2.2.34}$$

In particular, $\int |d\phi|^2$ is scaling invariant in dimension 2; in fact, this integral is even conformally invariant in dimension 2, as we shall explore below. In other dimensions, it becomes invariant only after a rescaling of the field ϕ according to (2.2.34). In order to compensate for the different scaling laws for $\int_{\mathbb{R}^n} |d\phi|^2 d^n x$ and $\int_{\mathbb{R}^n} |\phi(x)|^q d^n x$, one may also introduce coupling constants that are then scaled appropriately. To see this, let us consider

$$\int_{\mathbb{R}^n} (|d\phi|^2 - m^2 |\phi(x)|^2 + g_1 |\phi(x)|^3 + g_2 |\phi(x)|^4) d^n x; \tag{2.2.35}$$

here, m is the mass of the field as before. In order that this integral be scaling invariant, because of (2.2.29), each such polynomial term has to scale with a factor λ^n. If now ϕ scales according to (2.2.34), this leads to

$$m \to \lambda m, \qquad g_1 \to \lambda^{\frac{6-n}{2}} g_1, \qquad g_2 \to \lambda^{\frac{4-n}{2}} g_2. \tag{2.2.36}$$

Of course, this just re-expresses our discussion of (2.2.32), (2.2.33). For instance, in dimension ≤ 4, the integral of the polynomial ϕ^4 is controlled by that of $|d\phi|^2$, and therefore, we can afford a nonnegative scaling exponent for the coupling constant g_2. In general, an interaction term is called perturbatively renormalizable when the coupling constant scales with exponent 0, and superrenormalizable when it scales with a positive exponent.

Similarly, when we consider a term

$$\int_{\mathbb{R}^n} \gamma |\phi|^q |d\phi|^2 d^n x, \qquad (2.2.37)$$

we see that for $q > 0$, the coupling constant γ scales with exponent $\frac{(2-n)q}{2}$ which is negative for dimension > 2. Thus, such an interaction is renormalizable only for $n = 2$, but not for $n > 2$. This applies to the nonlinear sigma model discussed below (see (2.4.27), (2.4.28)),

$$\int_{\mathbb{R}^n} g_{ij}(\phi(x)) \frac{\partial \phi^i(x)}{\partial x^\alpha} \frac{\partial \phi^j(x)}{\partial x^\alpha} d^n x \qquad (2.2.38)$$

where g_{ij} denotes the metric tensor of the target N. When we expand (in normal coordinates)

$$g_{ij}(\phi) = \delta_{ij} + g_{ij,kl}\phi^k \phi^l + \text{higher-order terms}, \qquad (2.2.39)$$

we see that this model is not renormalizable for $n > 2$.

Next, if we have a Dirac term for a spinor field as in (2.2.9) in the Lagrangian, that is, if we have an action of the form

$$\int_{\mathbb{R}^n} i\langle \bar{\psi}, \slashed{D}\psi \rangle d^n x, \qquad (2.2.40)$$

we need the scaling behavior

$$\psi \to \lambda^{\frac{n-1}{2}} \psi \qquad (2.2.41)$$

to make it invariant. When we then have a mass term

$$m\bar{\psi}\psi, \qquad (2.2.42)$$

we obtain once more the scaling law $m \to \lambda m$ as in (2.2.36).

We finally consider gauge fields as in (2.2.27),

$$D_A = \partial + qA, \qquad (2.2.43)$$

with curvature $F = qdA + q^2 A \wedge A$, by (1.2.22). We can then expand the Lagrangian action (2.2.21) (in shorthand notation, as we are only interested in the growth orders),

$$\frac{1}{4q^2} \int_{\mathbb{R}^n} \|F\|^2 d^n x = \frac{1}{4} \int_{\mathbb{R}^n} ((dA)^2 + qA^2 dA + q^2 A^4) d^n x. \qquad (2.2.44)$$

As above, from the first term we get the scaling law (2.2.34)

$$A \rightarrow \lambda^{\frac{n-2}{2}} A, \qquad (2.2.45)$$

which then, taking always (2.2.29) into account, yields

$$q \rightarrow \lambda^{\frac{4-n}{2}} q. \qquad (2.2.46)$$

Thus, a gauge field Lagrangian action is renormalizable for dimension 4, but not above. The reader will then check that the same applies to the full interaction Lagrangians (2.2.21) and (2.2.26). Thus, we see that the Lagrangian action defined by (2.2.27) is indeed perturbatively renormalizable in dimension 4.

In contrast to this, let us consider the Einstein–Hilbert functional (1.1.163) of general relativity,

$$\frac{1}{16\pi G} \int_{\mathbb{R}^n} R(g) d^n x \qquad (2.2.47)$$

for the scalar curvature R of the metric g. Here, we have introduced the factor $\frac{1}{16\pi G}$ that we had neglected in the discussion of (1.1.163) above, G being Newton's gravitational constant. Also, we write $d^n x$ for the volume form because we expand around the flat metric,

$$g_{ij} = \delta_{ij} + \gamma_{ij} \sqrt{G}. \qquad (2.2.48)$$

We then obtain, with a similar shorthand notation as above, for the Einstein–Hilbert action (2.2.47)

$$\frac{1}{16\pi} \int_{\mathbb{R}^n} \left((dh)^2 + \sqrt{G} h (dh)^2 + \text{ higher-order terms} \right) d^n x, \qquad (2.2.49)$$

whence the scaling behavior $h \rightarrow \lambda^{\frac{n-2}{2}} h$ as before, and then

$$G \rightarrow \lambda^{2-n} G. \qquad (2.2.50)$$

Thus, the Lagrangian action of general relativity is renormalizable only in dimension 2, but not in dimension 4. This indicates that there should be difficulties in unifying gravity in dimension 4 with the other forces that are governed by a renormalizable Lagrangian of the form (2.2.27).

Here, we have only discussed perturbative renormalization (using [106]), but not nonperturbative renormalization, which is a more difficult issue. Some references for renormalization theory are [52, 113].

2.2.3 Elementary Particle Physics and the Standard Model

We now interrupt the process of setting up mathematical structures to discuss how this relates to elementary particles and in particular to their the contemporary theory as incorporated in the so-called standard model and its extensions. Physicists will, of course, know all this and may skip this section.

There exist four known basic physical forces: the electromagnetic, weak and strong forces and gravity. The standard model includes the first three of them, but leaves out gravity. In fact, it is the fundamental challenge of high-energy theoretical physics to construct a unified theory of all known forces, including gravity.

In any case, in a relativistic theory of elementary particles without gravitational effects, the Lagrangian should be invariant under the action of the Poincaré group or the double cover $G := Sl(2, \mathbb{C}) \ltimes \mathbb{R}^{1,3}$, see Sect. 1.3.4. This was most clearly formulated by Wigner who identified an elementary particle with an irreducible unitary representation of G satisfying certain physical restrictions, like $m^2 \geq 0$, where m is the mass. This principle is still fundamental, with the modification that one needs to consider groups that are larger than G, in order to account for internal symmetries of the particles beyond the spin. The principal for identifying that group combines the mathematical theory of group representations with scattering experiments designed to break the symmetry. To take an example from quantum mechanics, the Hamiltonian of a particle in a rotationally symmetric potential, $H = \frac{p^2}{2m} + V(|x|)$ (see (2.1.9)), commutes with the angular momentum operator $x \times p = \frac{\hbar}{i}(x \times \nabla)$, and therefore its eigenvalues, the energy levels, are degenerate. When an external magnetic field is applied, this symmetry gets broken and the energy levels, that is, the eigenvalues of the Hamiltonian, become distinct (Stern–Gerlach experiment, Zeeman effect). A further splitting of the energy levels of an electron in the presence of an external field is caused by its spin.

When it became clear that the proton and the neutron were very similar, except for their electrical charge, Heisenberg suggested that there was a single underlying particle, the nucleon, with a so-called isotopic spin, for short isospin, symmetry that was broken in the presence of electromagnetic interactions. This should correspond to the $L = \frac{1}{2}$ representation of $SU(2)$, as described at the end of Sect. 1.3.4. The proton and the neutron should correspond to the eigenvalues $\frac{1}{2}$ and $-\frac{1}{2}$ of $h = -it_3$. The subsequently discovered pions π^+, π^0, π^- should likewise correspond to the representation for $L = 1$ with the eigenvalues $1, 0, -1$ of h. This was supported by pion–nucleon scattering experiments. In those scattering experiments, the total charge Q as well as the total baryon number B was conserved (proton and neuron have $B = 1$, the pions have $B = 0$). In order to also incorporate the decay properties of other particles, Gell-Mann and Nishijima introduced another quantum number S, called strangeness, to be also preserved. There is a fundamental relation between the preceding numbers

$$Q = h + \frac{1}{2}(B + S). \tag{2.2.51}$$

Gell-Mann and Ne'eman then interpreted this as a consequence of the embedding of the isospin and the "hypercharge" $B + S$ into the larger Lie group $SU(3)$ of "flavor" symmetry, and this led Gell-Mann to suggest the existence of "quarks", particles corresponding to the basic representation of $SU(3)$ on $\mathbb{C}^3$. Finally, "color", another internal $SU(3)$ degree of freedom, was proposed. The modern theory of the strong interaction was then called quantum chromodynamics, after the Greek word for color.

We now turn to the unification of the fundamental forces. Electromagnetic and weak interactions (responsible for certain decay processes, like the beta decay of neutrons in nuclei) had been unified earlier, in 1967, in the so-called electroweak theory developed by Glashow, Salam and Weinberg, and in the early 1970s, the standard model combined this theory with quantum chromodynamics, the theory of strong interactions between quarks developed by Gell-Mann, Zweig and others. There are two types of particles in the theory: the fermions, which represent matter, and the bosons, which transmit forces between the fermions. Fermions have half-integer spin and satisfy the Pauli exclusion principle which states that two fermions cannot be in identical quantum states. This aspect will subsequently be incorporated into the formal framework by letting the fermions be odd Grassmann-valued, that is, anticommuting. Bosons have integer spins and do not have to satisfy the Pauli principle, and they will therefore be even Grassmann-valued, that is, commuting. The Lagrangian then has to couple the fermions and the bosons. There are four categories of bosons in the model: the photon that mediates electromagnetic interaction, the $W^{\pm}$ and Z boson for the weak force, eight types of gluons for the strong nuclear force, and finally the Higgs boson that induces spontaneous symmetry breaking of the gauge group for the electroweak interactions by a mechanism described in Sect. 2.2.4 below and that thereby provides masses to particles. While all the other bosons have been experimentally confirmed, with several of them predicted by the theory before their experimental observation, the Higgs boson has not yet been detected, but it may be detected soon with more powerful particle accelerators, because it should be seen at the energy scale where the unification between the electromagnetic and weak forces takes place (this is about 10^{12} electron volts, or about 10^{-16} times the Planck scale). Except for the Higgs boson and the W and Z bosons (which, in contrast to the photon, are massive, by the Higgs mechanism), the bosons are gauge particles, meaning that their contribution to the Lagrangian is invariant under gauge transformations from some internal symmetry group, as described in Sect. 2.2.1. The Lagrangians for bosons are of Yang–Mills type, as described in Sect. 1.2.3. The gauge group of electromagnetism is the Abelian group $U(1)$, as already explained. For the electroweak interaction, the gauge group is $SU(2) \times U(1)$. We should note, however, that the $SU(2)$ here is not the symmetry group of the weak interaction, which is not a gauge theory anyway. In fact, below the energy for the unification of the weak and the electromagnetic interactions, $SU(2)$ does not represent any symmetries. The gauge group $U(1)$ of electromagnetism is not the $U(1)$-factor in $SU(2) \times U(1)$, but rather a combination of the $U(1)$-subgroup of $SU(2)$ and the $U(1)$-factor in $SU(2) \times U(1)$ (Weinberg angle). For the strong interaction, the gauge group is $SU(3)$. The gauge group of the standard model is therefore the product $SU(3) \times SU(2) \times U(1)$. A class of extensions of the standard model, the so-called grand unified theories (GUTs), postulate that these groups are subgroups of a single large symmetry group, for example $SU(5)$. The grand unified theories have the advantage that they reduce the number of free parameters in the theory, see the remark at the end of Sect. 2.2.1. This symmetry, however, is only present at very high energies, but is reduced to $SU(3) \times SU(2) \times U(1)$ at lower energies (including those achievable by current particle accelerators) by a process of spontaneous symmetry breaking, see Sect. 2.2.4 below. Most of these grand unified theories, including $SU(5)$, had to be given up because (in contrast to the standard model) they predicted proton decay at a rate not observed in nature.

Supersymmetry, to be discussed below, postulates an additional symmetry between bosonic and fermionic particles.

For the fermions, we have 12 different types ("flavors", each representing a particle and its antiparticle) in the standard model. That flavor is changed by the weak interaction, mediated by the heavy W and Z gauge bosons. The fermions come in two classes, leptons (including the electron and the electron neutrino) and quarks. Only the latter ones participate in strong interactions, by a property called "color", as already mentioned above. Since, in contrast to the electroweak forces, the strong force grows with the distance between quarks, they become confined in hadrons, colorless combinations. These can consist either of three quarks, like the protons and neutrons, and therefore be fermionic (baryons), or of a quark–antiquark pair, and then be bosonic (mesons), like the pions. In particular, these particles, protons, neutrons, pions and so on, are not elementary, but composite. The fermions are also classified into three generations. Each fermion in one generation has counterparts in the other generations that only differ in their masses, for example the electron, the muon and the tau lepton. Ordinary matter consists of fermions of the first generation, that is, the electron, the electron neutrino and the up and down quarks, as the ones in the other generations are substantially more massive and quickly decay into lower-generation ones.

While the standard model is well confirmed (with some revision to account for the experimentally observed neutrino masses that had not been predicted by the original model), renormalizable and generally accepted, it cannot yet be the ultimate answer because it does not include gravity and does not fare well at the cosmological level. Also, it is not entirely satisfactory because it contains too many free parameters that are not theoretically derived, but can only be experimentally determined. (As mentioned, the number of these free parameters is reduced in the grand unified theories.)

The concepts behind the standard model, however, are theoretically very appealing. When extended by the more recent ideas of superstring theory, they may well lead to a general theory of all known physical forces, at least according to the present opinion of many, if not most, theoretical physicists.[12]

In the present book, we are not concerned with the detailed physical aspects of the standard model, but only with the underlying mathematical concepts. Therefore, we shall essentially only treat a toy model, the so-called sigma model, that itself does not pretend to describe actual physics, but which on the one hand exhibits many of the conceptual issues in a particularly transparent manner, and on the other hand

---

[12] As opinions in theoretical physics can change rapidly, this statement may no longer be up to date when this book goes to print, and perhaps not even at the time of writing. There seems to be at least a tendency towards growing scepticism with regard to the prospects of superstring theory.

constitutes the starting point for string theory which, in contrast, aims at physically valid predictions.

2.2.4 The Higgs Mechanism

For a d-component scalar field $\phi = (\phi^1, \ldots, \phi^d)$, we may consider the Lagrangian

$$F = \frac{1}{2} \partial_\mu \phi^i \, \partial^\mu \phi_i - \frac{1}{2} a_i^j \phi^i \phi_j \quad \text{(w.l.o.g. } a_i^j = a_j^i\text{).}$$

(Since the metric δ_{ij} on d-dimensional Euclidean space is flat, we can freely move indices up and down to conform to the usual summation conventions.)

By diagonalizing the quadratic form $a_i^j \phi^i \phi_j$ by an orthogonal transformation—which leaves $\partial_\mu \phi^i \, \partial^\mu \phi_i$ invariant—we can bring F into the form

$$F = \frac{1}{2} \partial_\mu \phi^i \, \partial^\mu \phi_i - \frac{1}{2} \sum_i \mu_i \phi^i \phi_i$$

(the μ_i are the eigenvalues of the symmetric matrix (a_{ij})).

If all $\mu_i \geq 0$, this Lagrangian describes d scalar particles of masses $m_i = \sqrt{\mu_i}$. Such an interpretation is no longer possible for negative μ_i.

More generally, for a multicomponent scalar field, we may consider the Lagrangian

$$F = \frac{1}{2} \partial_\mu \phi^i \, \partial^\mu \phi_i - V(\phi)$$

where the potential $V(\phi)$ incorporates self-interactions. Typically, V contains a quadratic term $a_{ij} \phi^i \phi^j$ and a higher-order term.

The classical vacuum corresponds to the minimum of V. The problem of symmetry breaking arises, namely that while V itself is invariant, the vacuum may not be invariant under the full symmetry group G of F. In that case, the vacuum consists of a whole G orbit, i.e., is degenerate.

Let us consider a simple example: ϕ is a real scalar field,

$$V(\phi) = \frac{1}{2} \mu \phi^2 + \frac{\lambda}{4} \phi^4.$$

In order to make V bounded from below, we assume $\lambda > 0$. If $\mu \geq 0$, then $\phi = 0$ is the vacuum, and the term $\frac{\lambda}{4} \phi^4$ may simply be treated as a higher-order perturbation for the Lagrangian

$$F_0 = \frac{1}{2} \partial_\mu \phi \partial^\mu \phi - \frac{1}{2} m^2 \phi^2 \quad (m = \sqrt{\mu})$$

for a scalar particle of mass m. If $\mu < 0$, the situation changes, and this interpretation is no longer possible. The vacuum is now located at

$$\phi_0 = v = \pm\sqrt{\frac{-\mu}{\lambda}}.$$

In order to make a perturbation around the vacuum, one now has to consider the shifted field

$$\tilde{\phi} = \phi - v,$$

which breaks the symmetry between ϕ and $-\phi$. In terms of $\tilde{\phi}$, the Lagrangian becomes

$$F = \frac{1}{2}\partial_\mu\tilde{\phi}\partial^\mu\tilde{\phi} + \frac{1}{2}\mu\tilde{\phi}^2 - \lambda v\tilde{\phi}^3 - \frac{\lambda}{4}\tilde{\phi}^4 \quad (+\text{an irrelevant constant term}).$$

Since μ is negative, we can interpret $\tilde{\phi}$ as a scalar particle of mass $m = \sqrt{-\mu}$.

We now apply a similar analysis for a massive complex scalar particle coupled with a gauge field (i.e., a massless vector particle) and consider the Lagrangian (cf. (2.2.1), (2.2.14), (2.2.17))

$$F = \frac{1}{2}(\partial_\mu\phi + A_\mu\phi)(\partial^\mu\phi^* - A^\mu\phi^*) - \lambda(|\phi|^2 - \tau^2)^2 - \frac{1}{4e^2}F_{\mu\nu}F^{\mu\nu} \quad (\tau > 0).$$
$$(2.2.52)$$

(The minus sign in front of A^μ arises because A^μ is in $\mathfrak{u}(1) \cong i\mathbb{R}$, and we have to take the complex conjugate $(A^\mu\phi)^* = -A^\mu\phi^*$.)

For $\lambda > 0$, the vacuum now is at

$$|\phi| = \tau,$$

i.e.,

$$\phi = \tau e^{i\vartheta}.$$

Thus, the vacuum is a nontrivial $U(1)$ orbit, i.e., degenerate. We therefore impose an additional gauge condition

$$\operatorname{Im}\phi = 0, \qquad \operatorname{Re}\phi > 0$$

which uniquely locates the vacuum at

$$\phi = \tau.$$

Again, we want to expand around the vacuum and put

$$\tilde{\phi} = \phi - \tau.$$

The Lagrangian becomes (up to a constant term)

$$F = \frac{1}{2}\partial_\mu\tilde{\phi}\partial^\mu\tilde{\phi} - \lambda\tilde{\phi}^2(\tilde{\phi} + 2\tau)^2 + \frac{1}{2}A_\mu A^\mu(\tilde{\phi} + \tau)^2 - \frac{1}{4e^2}F_{\mu\nu}F^{\mu\nu}$$

$$= \frac{1}{2}\partial_\mu\tilde{\phi}\partial^\mu\tilde{\phi} - 4\lambda\tau^2\tilde{\phi}^2 + \frac{1}{2}\tau^2 A_\mu A^\mu - \frac{1}{4e^2}F_{\mu\nu}F^{\mu\nu} + \text{higher-order terms.}$$

Thus, up to these higher-order terms, F describes a scalar particle $\tilde{\phi}$ of mass $m = 2\tau\sqrt{2\lambda}$ and a vector particle A of mass τ. Because of our above gauge condition, the scalar particle now has only one real degree of freedom left; the other degree of freedom has been gauged into the vector particle that has acquired a mass.

Alternatively, we write

$$\phi = e^{i\frac{\eta}{\tau}}(\tau + \xi) = \tau + \xi + i\eta + \mathcal{O}(\xi^2 + \eta^2)$$

and consider

$$\xi = e^{-i\frac{\eta}{\tau}}\phi - \tau,$$

$$A'_\mu = A_\mu - \frac{1}{\tau}\partial_\mu\eta.$$

Then F becomes

$$F = \frac{1}{2}\partial_\mu\xi\partial^\mu\xi - \frac{1}{2}\tau^2 A'_\mu A'^\mu - 4\lambda\tau^2\xi^2 - \frac{1}{4e^2}F_{\mu\nu}F^{\mu\nu}$$

$$+ \text{higher-order terms in } \xi \text{ and } A',$$

but η has disappeared, gauged into the vector particle A' that has acquired a mass.

Let us discuss the Higgs mechanism in more generality. Again, we start with a simple scalar field ϕ and a Lagrangian

$$F(\phi) = \frac{1}{2}\partial_\mu\phi\partial^\mu\phi - V(\phi).$$

We assume that ϕ takes values in a vector bundle with structure group G, and that V is G-invariant.

Let v be a classical vacuum, i.e., a minimizer of the potential V. Then

$$\frac{dV}{d\phi} = 0 \quad \text{at } \phi = v.$$

We assume that the symmetry group G is broken in the sense that v is only invariant under a smaller group $H \subset G$, but not under all of G.

We choose generators $\vartheta^1, \ldots, \vartheta^N$ of the Lie algebra $\mathfrak{g}$ of G in such a manner that $\vartheta^1, \ldots, \vartheta^M$ generate the Lie algebra $\mathfrak{h}$ of H. Thus

$$\vartheta^i v = 0 \quad \text{for } i = 1, \ldots, M,$$

$$\vartheta^j v \neq 0 \quad \text{for } j = M+1, \ldots, N.$$

Since V is G-invariant, the derivative of V in the directions tangent to the G-orbit of ϕ vanishes, that is,

$$\frac{dV}{d\phi}(\vartheta^j\phi) = \frac{d}{dt}V(\exp(t\vartheta^j)\phi)|_{t=0} = 0 \quad \text{for all } j.$$

Differentiating this relation w.r.t. ϕ yields

$$\frac{d^2V}{d\phi^2}\vartheta^j\phi + \frac{dV}{d\phi}\vartheta^j = 0 \quad \text{for all } j.$$

At $\phi = v$, $\frac{dV}{d\phi} = 0$, and hence

$$\frac{d^2V(v)}{d\phi^2}\vartheta^j v = 0.$$

Since for $j = M+1, \ldots, N$, $\vartheta^j v \neq 0$, $\vartheta^j v$ is an eigenvector of the Hessian $\frac{d^2V(v)}{d\phi^2}$ with zero eigenvalue. Since this Hessian gives the quadratic term in the expansion of our Lagrangian $F(\phi)$ at the vacuum $\phi = v$, and since the eigenvalues of this quadratic form are interpreted as squared masses, we interpret $\exp(\vartheta^j)v$ for $j = M+1, \ldots, N$ as a massless boson.

Thus, for each broken generator of the symmetry group, we have found a so-called massless Goldstone boson. To get the Higgs mechanism, we introduce a gauge field A, with values in $\mathfrak{g}$, that couples to our scalar field ϕ and consider the Lagrangian

$$F(\phi, A) = -\frac{1}{4}F^i_{\mu\nu}F_i^{\mu\nu} + \frac{1}{2}(\partial_\mu + A_\mu\phi)(\partial^\mu + A^\mu\phi) - V(\phi),$$

with a G-invariant potential V as before.

Again, we assume that the vacuum v is only H-invariant. We expand

$$\phi = v + \sum_{i=1}^{M}\xi_i\vartheta^i + \sum_{j=M+1}^{N}\eta_j\vartheta^j v + O(\xi^2 + \eta^2)$$

$$= \left\{\exp\left(\sum_{j=M+1}^{N}\eta_j\vartheta^j\right)\right\}\left(v + \sum_{i=1}^{M}\xi_i\vartheta^i\right).$$

With this expansion and with the notation $A_\mu = A_{i\mu}\vartheta^i$, we obtain the term

$$\text{Tr}\,\vartheta^i v\vartheta_j v A_{i\mu}A^{j\mu}$$

in $F(\phi, A)$, which we interpret as the mass term for the vector fields. The gauge change

$$A_{j'\mu} = A_{j\mu} - \partial_\mu\eta_j,$$

together with the fact that $\vartheta^j v$ for $j = M + 1, \ldots, N$ is a 0-eigenvector of the Hessian of V at v, makes the η^j disappear in the expansion of $F(\phi, A)$ up to second-order. Thus, we obtain $N - M$ massive vector fields that have absorbed the $N - M$ Goldstone bosons.

The idea of a gauge theoretic interpretation of spontaneous symmetry breaking was conceived independently by Englert and Brout and by Higgs in the early 1960s.

2.2.5 Supersymmetric Point Particles

We now take a step backwards and consider a one-dimensional domain. In geometric terms, the aim of this section is to derive a supersymmetric version of the action functional (1.1.119) for geodesics, discussed in Sect. 1.1.4. In physical terms, we want to introduce Lagrangians that exhibit a symmetry between bosonic and fermionic fields. The supersymmetric point particle is the simplest instance of this.

We start with the Euclidean case. We consider $(t, \theta) \in \mathbb{R}^{1|1}$ as coordinates (see Sect. 1.5.2), as well as scalar superfields

$$X^a(t, \theta) = \phi^a(t) + \psi^a(t)\theta, \quad a = 1, \ldots, d, \tag{2.2.53}$$

with ϕ^a even and ψ^a odd.[13] Our Lagrangian is

$$L_1 = \frac{1}{2}(\dot{\phi}^a \dot{\phi}_a + \psi^a \dot{\psi}_a) \tag{2.2.54}$$

and the action is

$$S_1 = \frac{1}{2} \int (\dot{\phi}^a \dot{\phi}_a + \psi^a \dot{\psi}_a) dt. \tag{2.2.55}$$

Here, t and θ are the independent variables, ϕ and ψ the dependent ones. ϕ is called a bosonic field, or boson for short, ψ a fermionic one or fermion.

Thus, both the arguments and the values of X are Grassmannian. Here, this makes X even. We also note that it is important that ψ be anticommuting in (2.2.55), as otherwise an integration by parts would imply that $\int \psi^a \dot{\psi}_a \, dt$ vanishes identically.

Remark

1. In the physics literature, one usually puts a factor i in front of the term $\psi^a \dot{\psi}_a$ in the Lagrangian F_1, and likewise in the other supersymmetric Lagrangians we shall treat here. Since the expression is Grassmann valued, this becomes a matter of convention, in contrast to the real-valued situation of (2.2.12). The convention with the i is compatible with the following convention usually adopted in the

---

[13] As ψ and θ are both odd, they anticommute. Otherwise, at this point, they have nothing to do with each other.

physics literature for defining a complex conjugation * on Grassmann variables. This conjugation should satisfy

$$(\tau_1 + \tau_2)^* = \tau_1^* + \tau_2^*, \qquad (\tau_1\tau_2)^* = \tau_2^*\tau_1^*. \qquad (2.2.56)$$

One defines the complex conjugate of an ordinary complex number as the ordinary complex conjugate, and one assumes that the generators $\vartheta_1, \ldots, \vartheta_N$ of the Grassmann algebra are real, i.e.,

$$\vartheta_i^* = \vartheta_i. \qquad (2.2.57)$$

Thus

$$(\vartheta_{\alpha_1}\vartheta_{\alpha_2}\cdots\vartheta_{\alpha_k})^* = \vartheta_{\alpha_k}\cdots\vartheta_{\alpha_2}\vartheta_{\alpha_1}. \qquad (2.2.58)$$

The elements of the Grassmann algebra are called supernumbers. A supernumber τ is then called real if $\tau^* = \tau$, imaginary if $\tau^* = -\tau$. Thus, $\vartheta_{\alpha_1}\cdots\vartheta_{\alpha_k}$ is real if $\frac{1}{2}k(k-1)$ is even, imaginary if $\frac{1}{2}k(k-1)$ is odd. With this convention, the term $\psi^a\dot{\psi}^a$, being the product of two real odd quantities, is purely imaginary, and the factor i then serves to make it real. In any case, a factor i in the Lagrangian in front of the ψ term would then require also a compensating factor in the supersymmetry transformations (2.2.68) below.
2. We may, in fact, put any factor κ in front of the term $\psi^a\dot{\psi}^a$ in the Lagrangian F_1 (and likewise, we may put factors in front of other terms we shall add to our Lagrangians to make them supersymmetric). We then simply need to compensate for this in our variations (2.2.68) below, for example by inserting a factor $1/\kappa$ into the right-hand side of the variation for ψ. With such a factor κ, we can then perform expansions of the Lagrangian and other quantities in terms of κ which is a useful device often seen in physics texts.

The Euler–Lagrange equations for L_1 are

$$\ddot{\phi}^a = 0, \qquad (2.2.59)$$
$$\dot{\psi}^a = 0, \qquad (2.2.60)$$

and L_1 describes a free superpoint particle. a is a vector index, and the setting can be generalized to particles moving on a Riemannian manifold M with metric $g_{ab}(y)dy^a \otimes dy^b$. $\phi(t)$ then transforms as a tangent vector to M, and one postulates that $\psi(t)$ likewise transforms as a tangent vector in $T_{\phi(t)}M$, i.e., under a change of coordinates $y = f(y')$ in M, one has

$$\psi^a = \frac{\partial y^a}{\partial y'^b}\psi'^b \quad (\psi \text{ is thus a vector with Grassmann coefficients}). \qquad (2.2.61)$$

(Note that here we are transforming the coordinates in the image; anticipating the discussion below, this is perfectly compatible with ψ being a spinor field, because this refers to the transformation behavior w.r.t. the independent variables.) Thus, ψ is an odd vector field along the map ϕ. In particular, the scalar product

$$\langle \dot{\phi}, \psi \rangle = g_{ab}(\phi)\dot{\phi}^a \psi^b \qquad (2.2.62)$$

is invariant under coordinate transformations on M. We may use the Lagrangian

$$L_2 := \frac{1}{2}g_{ab}(\phi)\dot{\phi}^a \dot{\phi}^b + \frac{1}{2}g_{ab}(\phi)\psi^a \nabla_{\frac{d}{dt}} \psi^b, \qquad (2.2.63)$$

with

$$\nabla_{\frac{d}{dt}} \psi^b = \frac{d}{dt}\psi^b + \dot{\phi}^a \Gamma^b_{ac}(\phi)\psi^c. \qquad (2.2.64)$$

The Euler–Lagrange equations for

$$S_2 = \int L_2(\phi, \psi)dt \qquad (2.2.65)$$

are

$$\nabla_{\frac{d}{dt}} \dot{\phi}^a - \frac{1}{2}R^a_{bcd}\dot{\phi}^b \psi^c \psi^d = 0, \qquad (2.2.66)$$

$$\nabla_{\frac{d}{dt}} \psi^a = 0. \qquad (2.2.67)$$

In contrast to (2.2.59), (2.2.60), these field equations couple ϕ and ψ. Equation (2.2.66) is the supersymmetric generalization of the geodesic equation $\nabla_{\frac{d}{dt}} \dot{x}^a = 0$, cf. (1.1.124), (1.1.125). When we consider ψ as an ordinary field, the equations (2.2.66), (2.2.67) describe a spinning particle in a gravitational field. Unless $\psi = 0$, the particle no longer moves along a geodesic, because of the presence of the second term in the first equation.

We return to the action S_1, and we perform the variations

$$\delta\phi^a = -\varepsilon\psi^a,$$
$$\delta\psi^a = \varepsilon\dot{\phi}^a \qquad (2.2.68)$$

with an odd parameter ε. The variation of the Lagrangian L_1 of S_1 is

$$\delta L_1 = \dot{\phi}^a \delta\dot{\phi}_a + \frac{1}{2}(\delta\psi^a)\dot{\psi}_a + \frac{1}{2}\psi^a \delta\dot{\psi}_a$$

$$= -\varepsilon\dot{\phi}^a \dot{\psi}_a + \frac{1}{2}\varepsilon\dot{\phi}^a \dot{\psi}_a + \frac{1}{2}\psi^a \varepsilon\ddot{\phi}_a$$

$$= -\varepsilon\dot{\phi}^a \dot{\psi}_a + \frac{1}{2}\varepsilon\dot{\phi}^a \dot{\psi}_a + \frac{1}{2}\frac{d}{dt}(\psi^a \varepsilon\dot{\phi}_a) - \frac{1}{2}\dot{\psi}^a \varepsilon\dot{\phi}_a$$

$$= -\varepsilon\frac{d}{dt}\left(\frac{1}{2}\psi^a \dot{\phi}_a\right) \quad \text{(using } \dot{\psi}^a\varepsilon = -\varepsilon\dot{\psi}^a, \text{ as } \varepsilon \text{ and } \psi^a \text{ are odd)}. \qquad (2.2.69)$$

Since δL_1 is a total derivative, it follows that S_1 is invariant under the variation (2.2.68). The point of the superspace formalism is now that this variation is induced by a variation of the independent variables t, θ. Namely, we consider the so-called supersymmetry generators

$$Q := \tau := \theta \partial_t + \partial_\theta, \tag{2.2.70}$$

$$P := \partial_t. \tag{2.2.71}$$

Here, Q and P are the notations usually employed in physics texts. The operator $Q = \tau$ should be compared with the operator $D := \partial_\theta - \theta \partial_t$ introduced in (1.5.34) in Sect. 1.5.2. Then

$$\varepsilon Q X^a(t, \theta) = \varepsilon Q(\phi^a(t) + \psi^a(t)\theta) = \varepsilon \dot{\phi}^a \theta - \varepsilon \psi^a$$

(since $\theta^2 = 0$ and θ commutes with $\dot{\phi}^a$, but anticommutes with ψ^a),   (2.2.72)

which yields (2.2.68). We have

$$[Q, Q] \equiv 2Q^2 = 2(\theta \partial_t + \partial_\theta)(\theta \partial_t + \partial_\theta) = 2\partial_t = 2P$$

(since, e.g., θ and ∂_θ anticommute).   (2.2.73)

Similarly

$$[D, D] = -2\partial_t = -2P, \tag{2.2.74}$$

$$[Q, P] = (\theta \partial_t + \partial_\theta)\partial_t - \partial_t(\theta \partial_t + \partial_\theta) = 0$$

(since ∂_θ anticommutes with θ and commutes with ∂_t),   (2.2.75)

$$[D, P] = 0, \tag{2.2.76}$$

$$[P, P] = 0. \tag{2.2.77}$$

Equations (2.2.73), (2.2.75), (2.2.77) mean that Q and P generate a super Lie algebra, see (1.5.5), i.e., a mod 2 graded vector space S over $\mathbb{C}$, endowed with a superbracket $[\cdot, \cdot]$ that is bilinear, mod 2 graded additive and superanticommutative, i.e.,

$$[A, B] = [B, A] \quad \text{if } A, B \text{ both are of odd degree,}$$

$$[A, B] = -[B, A] \quad \text{otherwise, i.e., if } A \text{ or } B \text{ is even}$$

and that satisfies the super-Jacobi identity (1.5.6),

$$(-1)^{ac}[A, [B, C]] + (-1)^{ba}[B, [C, A]] + (-1)^{cb}[C, [A, B]] = 0,$$

where a, b, c are the degrees of A, B and C, resp. Returning to (2.2.69), we have

$$\delta L_1 = -\varepsilon \dot{i}$$

with

$$I := \frac{1}{2}\psi^a \dot{\phi}^a.$$

We also have

$$\delta I = \frac{1}{2}\varepsilon \dot{\phi}^a \dot{\phi}^a + \frac{1}{2}\varepsilon \psi^a \dot{\psi}^a = \varepsilon L_1.$$

These variations for L_1 and I are quite similar to the ones for ϕ^a and ψ^a, compare (2.2.68), except that the roles of bosons and fermions have been exchanged. In the physics literature, the representation of the supersymmetry algebra on the ϕ^a, ψ^a space is called a "bosonic multiplet", whereas the one on the L_1, I space is called a "fermionic multiplet".

We are now going to consider a functional on $\mathbb{R}^{1|2}$ with coordinates (t, θ^1, θ^2), and the supersymmetry generators

$$Q_\alpha = \theta^\alpha \partial_t + \partial_{\theta^\alpha}, \tag{2.2.78}$$

$$P = \partial_t. \tag{2.2.79}$$

They span a super Lie algebra with

$$[Q_\alpha, Q_\beta] = 2\delta_{\alpha\beta} P,$$
$$[Q_\alpha, P] = 0 = [P, P]. \tag{2.2.80}$$

We try a superfield

$$X^a(t, \theta^\alpha) = \phi^a(t) + \psi^a_\alpha(t)\theta^\alpha. \tag{2.2.81}$$

We have

$$\varepsilon Q_1 X^a(t, \theta^1, \theta^2) = -\varepsilon \psi^a_1 + \varepsilon \dot{\phi}^a \theta^1 - \varepsilon \dot{\psi}^a_2 \theta^1 \theta^2, \tag{2.2.82}$$

and similarly for εQ_2. This is different from the previous situation as we now also get a $\theta^1\theta^2$ term that can neither be considered as a variation of ϕ^a nor as a variation of ψ^a. This problem stems from the fact that we now have only one bosonic field ϕ, but two fermionic fields ψ_1, ψ_2. Since supersymmetry mixes bosonic and fermionic fields, we need to introduce an additional bosonic field and consider the superfield

$$Y^a(t, \theta^\alpha) = \phi^a(t) + \psi^a_\alpha(t)\theta^\alpha + F^a(t)\theta^1\theta^2 \tag{2.2.83}$$

with an even F^a. We now get

$$\varepsilon Q_1 Y^a(t, \theta^\alpha) = -\varepsilon \psi^a_1 + \varepsilon \dot{\phi}^a \theta^1 + \varepsilon F^a \theta^2 - \varepsilon \dot{\psi}^a_2 \theta^1 \theta^2. \tag{2.2.84}$$

Thus, we get the variations

$$\delta\phi^a = -\varepsilon\psi_1^a,$$
$$\delta\psi_1^a = \varepsilon\dot{\phi}^a,$$
$$\delta\psi_2^a = \varepsilon F^a,$$
$$\delta F^a = -\varepsilon\dot{\psi}_2^a$$

(2.2.85)

for Q_1 and analogous variations for Q_2.

We consider the action

$$S_3 = \int \frac{1}{2}(\dot{\phi}^a\dot{\phi}_a + \psi_\alpha^a\dot{\psi}_{a\alpha} + F^a F_a)dt.$$

(2.2.86)

The Euler–Lagrange equations for L_3 are

$$\ddot{\phi}^a = 0,$$
$$\dot{\psi}_\alpha^a = 0,$$
$$F^a = 0.$$

(2.2.87)

Thus, the equations for the F^a are trivial, and the F^a are auxiliary variables that do not evolve and can be eliminated. They are only needed to close the supersymmetry algebra. We also observe that on-shell, i.e., if the equations of motion (2.2.87) are satisfied, we have

$$\varepsilon Q_1 X^a(t, \theta^\alpha) = -\varepsilon\psi_1^a + \varepsilon\dot{\phi}^a\theta^1$$

(2.2.88)

so that the supersymmetry algebra closes here without the F^a field. On-shell, the number of degrees of freedom of the ψ_α^a fields is reduced so that we no longer need the F^a field in order to restore the balance between bosonic and fermionic fields, and on-shell, F^a vanishes anyway.

We may also write things in a more invariant manner. Namely, with

$$X(t, \theta) = \phi(t) + \psi(t)\theta,$$

(2.2.89)

we have

$$DX = -\dot{\phi}\theta - \psi$$

(2.2.90)

and

$$D(DX) = -(\dot{\phi} + \dot{\psi}\theta),$$

(2.2.91)

and so we have

$$S_1 = \int \frac{1}{2}DXD(DX)dtd\theta.$$

(2.2.92)

Since the supersymmetry generator Q anticommutes with D, we now see directly, without any need for further computation, that L_1 remains invariant under super-symmetry transformations. Similarly, to represent S_3, we consider the operators

$$D_\alpha = -\theta^\alpha \partial_t + \partial_{\theta^\alpha}. \tag{2.2.93}$$

From (2.2.83), we obtain

$$S_3 = \frac{1}{4} \int \epsilon^{\alpha\beta} D_\alpha Y D_\beta Y \, dt \, d\theta^2 d\theta^1, \tag{2.2.94}$$

for the field Y, where the antisymmetric ϵ-tensor satisfies

$$\epsilon^{12} = -\epsilon^{21} = 1. \tag{2.2.95}$$

We observe that, in contrast to (2.2.92), (2.2.94) contains only two Ds, the reason being that here we have two odd variables, θ^1 and θ^2, that are integrated. The Euler–Lagrange equations (2.2.87) for L_3 then become

$$\epsilon^{\alpha\beta} D_\alpha D_\beta Y = 0. \tag{2.2.96}$$

We now wish to include self-interaction terms in the functional and consider a (smooth) potential function of the superfields Y^a,

$$W(Y^a) = W(\phi^a + \psi_\alpha^a \theta^\alpha + F^a \theta^1 \theta^2). \tag{2.2.97}$$

Thus, we have the expansion

$$W(Y) = w(\phi) + \frac{\partial w(\phi)}{\partial \phi^a}(\psi_\alpha^a \theta^\alpha + F^a \theta^1 \theta^2) + \frac{1}{2}\frac{\partial^2 w(\phi)}{\partial \phi^a \partial \phi^b}\psi_\alpha^a \theta^\alpha \psi_\beta^b \theta^\beta. \tag{2.2.98}$$

We introduce an interaction Lagrangian

$$L_{int} = -\int W(Y(t,\theta^1,\theta^2)) dt \, d\theta^2 d\theta^1 \tag{2.2.99}$$

$$= -\int \left(\frac{\partial w(\phi(t))}{\partial \phi^a(t)}F^a + \frac{\partial^2 w(\phi(t))}{\partial \phi^a(t)\partial \phi^b(t)}\psi_1^a \psi_2^b\right) dt. \tag{2.2.100}$$

The total Lagrangian is

$$L_3 + L_{int}. \tag{2.2.101}$$

The corresponding Euler–Lagrange equations include the following equation for F

$$F^a = \frac{\partial w(\phi)}{\partial \phi^a}. \tag{2.2.102}$$

Again, this is an algebraic equation and thus eliminates F^a, and we may write

$$L_3 + L_{int} = \int dt \left(\frac{1}{2}\dot\phi^a \dot\phi_a + \frac{1}{2}\psi_\alpha^a \dot\psi_{a\alpha} - \frac{1}{2}\frac{\partial w(\phi)}{\partial \phi^a}\frac{\partial w(\phi)}{\partial \phi_a} - \frac{\partial^2 w(\phi)}{\partial \phi^b \partial \phi^a}\psi_1^a \psi_2^b\right). \tag{2.2.103}$$

2.3 Variational Aspects

2.3.1 The Euler–Lagrange Equations

Here, we present a brief summary of the calculus of variations as needed for treating action functionals and their symmetries. For more details, we refer to [66]. We consider a Lagrangian

$$L = \int F(x, u(x), du(x))dx \tag{2.3.1}$$

and variations

$$u(x) \rightarrow u(x) + s\delta u(x); \tag{2.3.2}$$

here, s is a parameter, and $\delta u(x)$ is the variation of u at x.[14] This means the following:

$$\delta L(u)(\delta u) := \frac{\delta L(u)}{\delta u} := \frac{d}{ds} \int F(x, u(x) + s\delta u(x), d(u(x) + s\delta u(x)))dx_{|s=0}.$$
$$\tag{2.3.3}$$

More generally, one may consider a C^2-family of diffeomorphisms $h_s(u)$ of the dependent variables, defined for s in some neighborhood of 0, with h_0 being the identity, and

$$\frac{d}{ds} h_s(u(x))_{|s=0} = \delta u(x), \tag{2.3.4}$$

and

$$\delta L(u)(\delta u) = \frac{d}{ds} \int F(x, h_s(u(x)), d(h_s(u(x))))dx_{|s=0}. \tag{2.3.5}$$

Since we consider only infinitesimal variations, (2.3.3) and (2.3.5) are the same, and we may use either formulation.

We now assume that

$$\delta L(u)(\delta u) \left(= \frac{\delta L(u)}{\delta u} \right) = 0 \tag{2.3.6}$$

for a variation δu. We compute

$$\delta L(u)(\delta u) = \int \left(F_u(x, u(x), du(x)) \delta u(x) + F_{p^\alpha}(x, u(x), du(x)) \frac{\partial}{\partial x^\alpha} \delta u(x) \right) dx$$

---

[14]The integration is supposed to take place on some domain Ω, but as that domain will play no essential role, we suppress it in our notation. In many situations, the variations δu are required to satisfy certain conditions at the boundary of Ω (because u itself is constrained there), but again, that will not be essential in the present context. Of course, it is important to realize that the integral in the definition of the Lagrangian is a definite one. Likewise, we suppress other constraints that u may have to satisfy, and that need to be preserved by its variations.

$$= \int \left(F_u(x, u(x), du(x)) - \frac{d}{dx^\alpha} F_{p^\alpha}(x, u(x), du(x)) \right) \delta u(x) dx$$

$$(2.3.7)$$

where p^α is a dummy variable for the place where $\frac{\partial u}{\partial x^\alpha}$ is inserted, and subscripts denote partial derivatives, e.g. $F_u := \frac{\partial F}{\partial u}$. (In fact, u might be vector valued, and in that case F_u stands for all the partial derivatives of F w.r.t. the components of u.) For the last line, we have integrated by parts, assuming that the variation δu is such that no boundary term occurs. We note the full derivative $\frac{d}{dx^\alpha}$ that indicates that we need to differentiate $F_{p^\alpha}(x, u(x), du(x))$ for all three occurrences of x.

Comparing (2.3.7) with (2.1.52), we obtain

$$\frac{\delta L(u)}{\delta u(x)} = F_u(x, u(x), du(x)) - \frac{d}{dx^\alpha} F_{p^\alpha}(x, u(x), du(x)). \qquad (2.3.8)$$

Thus, $\frac{\delta L(u)}{\delta u(x)}$ represents the Euler–Lagrange operator.

We now assume that u is a stationary point, e.g., a minimizer of L, in the sense that (2.3.6) holds for all variations δu satisfying an appropriate boundary condition. We then obtain the **Euler–Lagrange equations**

$$\frac{\delta L(u)}{\delta u(x)} = F_u(x, u(x), du(x)) - \frac{d}{dx^\alpha} F_{p^\alpha}(x, u(x), du(x)) = 0. \qquad (2.3.9)$$

When we wish to derive things in a geometrically invariant way, we should change the preceding formalism slightly. The reason is that the integration measure dx employed in (2.3.1) is not geometrically invariant. More natural is the volume form

$$\sqrt{\det g_{ij}}dx \text{ for a Riemannian metric } g_{ij}. \qquad (2.3.10)$$

Thus, in place of (2.3.1), we should consider

$$L = \int G(x, u(x), du(x))\sqrt{\det g_{ij}}dx. \qquad (2.3.11)$$

We abbreviate

$$\sqrt{g} := \sqrt{\det g_{ij}}. \qquad (2.3.12)$$

The Euler–Lagrange equations (2.3.9) then become

$$\frac{\delta L(u)}{\delta u(x)} = G_u(x, u(x), du(x)) - \frac{1}{\sqrt{g}}\frac{d}{dx^\alpha}(\sqrt{g}G_{p^\alpha}(x, u(x), du(x))) = 0. \quad (2.3.13)$$

2.3.2 Symmetries and Invariances: Noether's Theorem

We consider an action

$$L = \int F(x, u(x), du(x))dx \qquad (2.3.14)$$

that is infinitesimally invariant under some variation

$$u(x) \to u(x) + s\eta(x). \qquad (2.3.15)$$

As just explained, the invariance means that

$$\delta L := \frac{d}{ds} \int F(x, u(x) + s\eta(x), d(u(x) + s\eta(x)))dx_{|s=0} = 0. \qquad (2.3.16)$$

In contrast to the preceding, here we consider arbitrary fields u, but only particular variations η—above, we had considered arbitrary variations δu for a particular u.

Again, we may alternatively consider a C^2-family of diffeomorphisms $h_s(u)$ of the dependent variables, defined for s in some neighborhood of 0, with h_0 being the identity, and $\frac{d}{ds}h_s(u(x))_{|s=0} = \eta(x)$. We now assume

$$\int F(x, h_s(u(x)), dh_s(u(x)))dx = \int F(x, u(x), du(x))dx$$

for all s near 0 and all admissible u. The interpretation that a variation arises from a diffeomorphism of the dependent variables that leaves the action invariant is useful when one wants to analyze invariances in the context of global analysis.

As in (2.3.7), we obtain

$$0 = \int \left(F_u(x, u(x), du(x))\eta(x) + F_{p^\alpha}(x, u(x), du(x))\frac{\partial}{\partial x^\alpha}\eta(x) \right)dx. \qquad (2.3.17)$$

We now consider a more general variation

$$u(x) \to u(x) + s(x)\eta(x), \qquad (2.3.18)$$

that is, where the variation parameter s may also depend on x. Since the variation δL vanishes for constant s, it must now be proportional to the derivative of s, that is,

$$\delta L = \int F_{p^\alpha}(x, u(x), du(x))\eta(x)\frac{\partial}{\partial x^\alpha}s(x)dx. \qquad (2.3.19)$$

If we now assume in addition that u is stationary, that is, δL vanishes for all variations, (2.3.19) vanishes as well, and we conclude

$$\frac{d}{dx^\alpha}(F_{p^\alpha}(x, u(x), du(x))\eta(x)) = 0. \qquad (2.3.20)$$

This is a special version of **Noether's theorem**. We define the Noether current

$$j^\alpha(x) := F_{p^\alpha}(x, u(x), du(x))\eta(x). \qquad (2.3.21)$$

Noether's theorem thus says that j is conserved in the sense that

$$\text{div } j = \frac{\partial}{\partial x^\alpha}j^\alpha(x) = 0. \qquad (2.3.22)$$

As at the end of Sect. 2.3.1, when we consider a functional of the form (2.3.10), we obtain the geometric version of Noether's theorem,

$$\text{div } j = \frac{1}{\sqrt{g}} \frac{\partial}{\partial x^\alpha} (\sqrt{g} j^\alpha(x)) = 0, \qquad (2.3.23)$$

with $j^\alpha(x) := G_{p^\alpha}(x, u(x), du(x)) \eta(x)$.

For the general version of Noether's theorem, we also allow for variations of the independent variable x. That means that we consider

$$x \to x' := x + s\delta x, \qquad (2.3.24)$$

$$u(x) \to u'(x') := \psi(u(x)) := u(x) + s\delta\psi(x). \qquad (2.3.25)$$

When we write

$$u'(x) = u(x) + s\eta(x) \qquad (2.3.26)$$

we have

$$\eta = \delta\psi - \frac{du}{dx^\beta} \delta x^\beta. \qquad (2.3.27)$$

Since now the integration measure dx also varies under (2.3.24), the Noether current becomes

$$j^\alpha := F_{p^\alpha}(x, u(x), du(x)) \left(\delta\psi - \frac{du}{dx^\beta} \delta x^\beta \right) + F(x, u(x), du(x)) \delta x^\alpha, \quad (2.3.28)$$

and again a conserved quantity,

$$\text{div } j = \frac{\partial}{\partial x^\alpha} j^\alpha(x) = 0. \qquad (2.3.29)$$

Here as well, in the Riemannian setting, we instead have

$$j^\alpha := G_{p^\alpha}(x, u(x), du(x)) \left(\delta\psi - \frac{du}{dx^\beta} \delta x^\beta \right) + G(x, u(x), du(x)) \delta x^\alpha \quad (2.3.30)$$

and (2.3.29) should be replaced by

$$\text{div } j = \frac{1}{\sqrt{g}} \frac{\partial}{\partial x^\alpha} (\sqrt{g} j^\alpha(x)) = 0, \qquad (2.3.31)$$

cf. (2.3.23).

Finally, in many situations, the Lagrangian is only invariant up to a divergence term, that is,

$$\delta L + \int \text{div } A(u, \delta x, \delta u) = 0. \qquad (2.3.32)$$

In this case, we obtain that (in abbreviated notation when compared to (2.3.28))

$$\operatorname{div}(F_p \delta u + (F - F_p du)\delta x + A(u, \delta x, \delta u)) = 0. \qquad (2.3.33)$$

The standard example is the conservation of energy for time-invariant Lagrangians. We consider the action

$$S_0 = \int \left(\frac{1}{2}\dot{\phi}^a \dot{\phi}_a - V(\phi) \right) dt. \qquad (2.3.34)$$

Since the integrand does not depend explicitly on t, it is invariant under a variation $t \to t + \delta t$, and from (2.3.28), we conclude that the negative of the Hamiltonian (= energy) is preserved, this being given by

$$\frac{1}{2}\dot{\phi}^a \dot{\phi}_a + V(\phi). \qquad (2.3.35)$$

The same happens for our supersymmetric Lagrangian (2.2.55),

$$S_1 = \frac{1}{2}\int (\dot{\phi}^a \dot{\phi}_a + \psi^a \dot{\psi}_a) dt; \qquad (2.3.36)$$

again, the Noether current is

$$-\frac{1}{2}\dot{\phi}^a \dot{\phi}_a, \qquad (2.3.37)$$

that is, minus the Hamiltonian. We observe that the Noether current here contains only the bosonic field ϕ, not the fermionic one ψ. We may also consider supersymmetry invariance in this framework. When we perform the variations (2.2.68), we compute that the associated current is given by

$$-\epsilon \dot{\phi}\psi. \qquad (2.3.38)$$

The superspace formalism represented the supersymmetry variation as a variation of the independent variables. Since S_1 in (2.2.92), however, also contains terms of the form $D^2 X$, the formalism needs to be slightly extended to carry over. The Lagrangian S_3 as written in (2.2.94) does not present this problem, and so, in that case, the conserved current can be computed from a variation of the independent variables without the need for an extension of the formalism, except that of course we now need to take the signs into account as always when performing supercomputations.

We now finally derive some implications of Noether's theorem in the Minkowski setting. According to (2.3.29), invariance implies a conserved current:

$$\partial_\alpha j^\alpha = 0. \qquad (2.3.39)$$

From this, we obtain a conserved charge:

$$Q = \int d^{d-1}x j^0, \qquad (2.3.40)$$

where j^0 is the time component of j and $d^{d-1}x$ denotes the integration over a space-like slice,

$$
\frac{d}{dt}Q = \int d^{d-1}x \; \partial_0 j^0
$$

$$
= -\int d^{d-1}x \; \partial_a \, j^a \quad (a \text{ running over spatial indices}) \text{ by (2.3.24)},
$$

$$
= 0 \tag{2.3.41}
$$

by the divergence theorem when j vanishes sufficiently quickly at spatial infinity.

2.4 The Sigma Model

In this section, we discuss an action functional that is fundamental to conformal field theory and string theory, the sigma model and its nonlinear and supersymmetric versions. In the mathematical literature, the corresponding theory appears under the name of harmonic maps, and we refer to [65] for a detailed treatment with proofs and references; for the supersymmetric version, the harmonic map needs to be coupled with a Dirac field as treated in [16]. A monograph about this topic from physics is [73].

2.4.1 The Linear Sigma Model

We let M be a Riemannian manifold of dimension m, with metric tensor in local coordinates $(\gamma_{\alpha\beta})_{\alpha,\beta=1,\dots,m}$.[15]
We recall the following notation:

$$
(\gamma^{\alpha\beta})_{\alpha,\beta=1,\dots,m} = (\gamma_{\alpha\beta})^{-1}_{\alpha,\beta} \quad \text{(inverse metric tensor)},
$$

$$
\gamma = \det(\gamma_{\alpha\beta}),
$$

$$
\Gamma^\alpha_{\beta\eta} = \frac{1}{2}\gamma^{\alpha\delta}(\gamma_{\beta\delta,\eta} + \gamma_{\eta\delta,\beta} - \gamma_{\beta\eta,\delta}) \quad \text{(Christoffel symbols)}.
$$

For a function $\phi : M \to \mathbb{R}$ of class C^1, we consider

$$
\gamma^{\alpha\beta}(x)\frac{\partial \phi(x)}{\partial x^\alpha}\frac{\partial \phi(x)}{\partial x^\beta} \tag{2.4.1}
$$

---

[15]The conventions here are different from the ones established in Sect. 1.1.1 where the metric of M was denoted by g_{ij}. The reason is that for the nonlinear sigma model, another manifold will come into play, the physical space(–time) whose metric will then be denoted by g_{ij}. M will be the world sheet instead.

in local coordinates $(x^1, \ldots, x^m)$ on M. The quantity (2.4.1) is simply the square of the norm of the differential $d\phi = \frac{\partial \phi}{\partial x^\alpha} dx^\alpha$, which is a section of the cotangent bundle T^*M, that is,

$$\gamma^{\alpha\beta}(x) \frac{\partial \phi(x)}{\partial x^\alpha} \frac{\partial \phi(x)}{\partial x^\beta} = \langle d\phi, d\phi \rangle_{T^*M} = \|d\phi\|^2 \qquad (2.4.2)$$

in hopefully self-explanatory notation. Therefore, it is clear that (2.4.1) is invariant under coordinate changes.

The Dirichlet integral of ϕ is then

$$S(\phi) = \frac{1}{2} \int_M \gamma^{\alpha\beta}(x) \frac{\partial \phi}{\partial x^\alpha} \frac{\partial \phi}{\partial x^\beta} \sqrt{\gamma} dx^1 \cdots dx^m = \frac{1}{2} \int_M \|d\phi\|^2 dvol_\gamma(M). \quad (2.4.3)$$

Minimizers are harmonic functions; they solve the Laplace–Beltrami equation (see (1.1.103)) (the Euler–Lagrange equation for $S(\phi)$)

$$\Delta_M \phi = \frac{1}{\sqrt{\gamma}} \frac{\partial}{\partial x^\alpha} \left(\sqrt{\gamma} \gamma^{\alpha\beta} \frac{\partial}{\partial x^\beta} \phi \right) = 0. \qquad (2.4.4)$$

We now specialize this to the case where M is two-dimensional, that is, a surface equipped with some Riemannian metric. According to the conventions set up in Sect. 1.1.2, we can then let the indices α, β stand for $z, \bar{z}$, where $z = x^1 + ix^2$ is a complex coordinate; when we want to avoid indices, we shall also write $z = x + iy$, as in Sect. 1.1.2. Thus

$$\|d\phi\|^2 = \gamma^{zz} \frac{\partial \phi}{\partial z} \frac{\partial \phi}{\partial z} + 2\gamma^{z\bar{z}} \frac{\partial \phi}{\partial z} \frac{\partial \phi}{\partial \bar{z}} + \gamma^{\bar{z}\bar{z}} \frac{\partial \phi}{\partial \bar{z}} \frac{\partial \phi}{\partial \bar{z}} \qquad (2.4.5)$$

and, recalling (1.1.76),

$$S(\phi) = \frac{1}{2} \int_M \gamma^{\alpha\beta} \frac{\partial \phi}{\partial x^\alpha} \frac{\partial \phi}{\partial x^\beta} \sqrt{\gamma_{11}\gamma_{22} - \gamma_{12}^2} \, dx \wedge dy$$

$$= \frac{1}{2} \int_M \left(\gamma^{zz} \frac{\partial \phi}{\partial z} \frac{\partial \phi}{\partial z} + 2\gamma^{z\bar{z}} \frac{\partial \phi}{\partial z} \frac{\partial \phi}{\partial \bar{z}} + \gamma^{\bar{z}\bar{z}} \frac{\partial \phi}{\partial \bar{z}} \frac{\partial \phi}{\partial \bar{z}} \right)$$

$$\times \sqrt{\gamma_{zz}\gamma_{\bar{z}\bar{z}} - \gamma_{z\bar{z}}^2} \, dz \wedge d\bar{z}. \qquad (2.4.6)$$

A fundamental point in the sequel will be to consider S not only as a function of the field ϕ, but also of the metric γ. We thus write

$$S(\phi, \gamma). \qquad (2.4.7)$$

Naturally, we then also want to study the effect of variations $\delta\gamma$ of the metric on S; it is in fact more convenient to study variations of the inverse metric γ^{-1}. Observing that $\sqrt{\gamma_{11}\gamma_{22} - \gamma_{12}^2} = (\sqrt{\gamma^{11}\gamma^{22} - (\gamma^{12})^2})^{-1}$, we compute

$$\delta S(\phi, \gamma) = \int T_{\alpha\beta} \delta\gamma^{\alpha\beta} \sqrt{\gamma_{11}\gamma_{22} - \gamma_{12}^2} \, dx \wedge dy \qquad (2.4.8)$$

with

$$T_{\alpha\beta} = \frac{\partial\phi}{\partial x^\alpha}\frac{\partial\phi}{\partial x^\beta} - \frac{1}{2}\gamma_{\alpha\beta}\gamma^{\epsilon\eta}\frac{\partial\phi}{\partial x^\epsilon}\frac{\partial\phi}{\partial x^\eta}. \tag{2.4.9}$$

We call $T_{\alpha\beta}$ the **energy–momentum tensor**[16] and observe that T is trace-free, that is,

$$T^\alpha_\alpha = 0. \tag{2.4.10}$$

Here, the dimension 2 is essential. The reason why T is trace free is the **conformal invariance** of S in dimension 2. This simply means that when we change the metric to $e^\sigma \gamma$ for some function $\sigma : M \to \mathbb{R}$, then S stays invariant:

$$S(\phi, e^\sigma\gamma) = S(\phi, \gamma). \tag{2.4.11}$$

Infinitesimally, the variation of γ^{-1} is $-\delta\sigma\,\gamma^{-1}$, and from (2.4.8), (2.4.11), we get

$$0 = \int T_{\alpha\beta}\,\delta\sigma\,\gamma^{\alpha\beta}\sqrt{\gamma_{11}\gamma_{22} - \gamma_{12}^2}\,dx \wedge dy$$
$$= \int T^\alpha_\alpha\,\delta\sigma\sqrt{\gamma_{11}\gamma_{22} - \gamma_{12}^2}\,dx \wedge dy \tag{2.4.12}$$

for all variations $\delta\sigma$, which implies (2.4.10).

We now consider this from a slightly different point of view. A Riemannian metric γ on a surface induces the structure of a Riemann surface Σ, as defined in Sect. 1.1.2, via the uniformization theorem (for a detailed treatment, we refer to [64]). As a consequence, we can find holomorphic coordinates $z = x + iy$ for which the metric is diagonal, that is,

$$\gamma_{12} = 0 \quad \text{and} \quad \gamma_{11} = \gamma_{22}, \tag{2.4.13}$$

or equivalently,

$$\gamma_{zz} = 0 = \gamma_{\bar{z}\bar{z}}. \tag{2.4.14}$$

We can then express the metric tensor by a single (nonvanishing) scalar function λ, that is, as

$$\lambda^2 dz d\bar{z}; \tag{2.4.15}$$

cf. (1.4.18). As explained in Sect. 1.4.2, a Riemann surface Σ can be considered as a conformal equivalence class of metrics of the form (2.4.15). When we choose

---

[16]Since we consider a Euclidean instead of a Minkowskian situation, we cannot distinguish between temporal and spatial directions, and thus also not between energy and momentum as quantities that are preserved because of temporal or spatial invariance, according to Noether's theorem. This explains the name energy–momentum tensor here. In general relativity, the energy–momentum tensor emerges because Lorentz invariance combines temporal and spatial invariance.

coordinates so that this holds, our functional S also simplifies:

$$S(\phi, \gamma) = \frac{1}{2} \int_M \left(\frac{\partial \phi}{\partial x} \frac{\partial \phi}{\partial x} + \frac{\partial \phi}{\partial y} \frac{\partial \phi}{\partial y} \right) dx \wedge dy = \int_M \frac{\partial \phi}{\partial z} \frac{\partial \phi}{\partial \bar{z}} \, i dz \wedge d\bar{z}. \quad (2.4.16)$$

Thus, the dependence on the metric disappears, except that the holomorphic coordinates $z = x + iy$ have been chosen so as to diagonalize the metric. In other words, S is a function of the equivalence class of metrics encoded by Σ, and we can write it as

$$S(\phi, \Sigma). \quad (2.4.17)$$

In these conformal coordinates, that is, where (2.4.13), (2.4.14) hold, the condition (2.4.10) becomes

$$T_{z\bar{z}} = 0. \quad (2.4.18)$$

Since we take the field ϕ to be real-valued here, we also have $\frac{\partial \phi}{\partial z} = \overline{\frac{\partial \phi}{\partial z}}$, and so $T_{\bar{z}\bar{z}} = T_{zz}$. Therefore T is determined by its component T_{zz}. Taking its the transformation behavior into account as well, the energy–momentum tensor becomes a quadratic differential

$$T_{zz} dz^2 = \left(\frac{\partial \phi}{\partial z} \right)^2 dz^2 \quad (2.4.19)$$

from (2.4.9).

When the metric takes the form (2.4.15), the Laplace–Beltrami equation satisfied by critical points of S also simplifies:

$$\frac{4}{\lambda^2} \frac{\partial^2 \phi}{\partial z \partial \bar{z}} = 0, \quad (2.4.20)$$

which is equivalent to the simpler equation

$$4 \frac{\partial^2 \phi}{\partial z \partial \bar{z}} = 0. \quad (2.4.21)$$

In this presentation, the dependence on the metric is no longer visible. This simply comes from the fact that we write the equation in local coordinates, and local coordinate neighborhoods are conformally equivalent to domains in the Euclidean complex line $\mathbb{C}$. When we return to the global aspects, as explained, we have a functional $S(\phi, \Sigma)$ that depends on the Riemann surface Σ. In Sect. 1.4.2, we have considered the moduli space of Riemann surfaces, and we can thus consider S as a functional on the moduli space M_p of Riemann surfaces of some given genus p. As explained in that section, that moduli space is not compact for $p > 0$, and one should then consider an extension of S to a compactification of M_p. Such a compactification was constructed by pinching homotopically nontrivial closed curves (represented by closed geodesics w.r.t. a hyperbolic metric for $p > 1$), thus creating surfaces with singularities. Those singularities were then removed by compactifying the resulting surfaces by two points, one for each side of the closed geodesic. This now connects well with the behavior of the functional S, because its critical points, the solutions of (2.4.20), (2.4.21), are harmonic functions. And bounded harmonic functions can be smoothly extended across isolated singularities. That means that if we consider a sequence of degenerating Riemann surfaces Σ_n and controlled harmonic functions u_n (with some suitable norm bounded independently of n) on them, we can pass to the limit (of some subsequence) that then defines a harmonic function u on the Riemann surface Σ obtained by the described compactification of the limit of the Riemann surfaces. That harmonic function is then smooth on all of Σ, and in particular, it does not feel the presence or the position of the puncture, that is, of the points added for the compactification. In particular, the functional S then naturally extends not only to the Deligne–Mumford compactification, but also to the Baily–Satake compactification $\overline{\overline{M}}_p$ (see Sect. 1.4.3 of the moduli space M_p). For more details, see [62].

There is one point here that will become important below in Sect. 2.5. While the equation of motion, our Euler–Lagrange equation (2.4.20), is conformally invariant in the sense that the conformal factor $\frac{1}{\lambda^2}$ plays no role, the corresponding differential operator, the Laplace–Beltrami operator $\frac{4}{\lambda^2}\frac{\partial^2}{\partial z \partial \bar{z}}$, is not conformally invariant itself.

From (2.4.20), we see directly that the energy–momentum tensor as given by (2.4.19) is holomorphic at a solution of (2.4.20):

$$\frac{\partial T_{zz}}{\partial \bar{z}} = 0. \tag{2.4.22}$$

In conclusion, the energy–momentum tensor yields a holomorphic quadratic differential $T_{zz}dz^2 = (\frac{\partial \phi}{\partial z})^2 dz^2$ on our Riemann surface Σ.

There is a deeper reason why T is holomorphic. As we shall now explain, S is invariant under diffeomorphisms, and by Noether's theorem, this yields a conserved current, that is, a divergence-free quantity. That latter equation then turns out to be equivalent to (2.4.22). The reason is simply that (2.4.6), or equivalently (2.4.16), is invariant under coordinate changes. In mathematical terms, as explained in Sect. 1.1.1, this means that we compose the field ϕ with a diffeomorphism h of our surface and simultaneously pull the metric γ in (2.4.6) or the area form $dx \wedge dy$ in (2.4.16) back by that diffeomorphism. In other words, we have

$$S(\phi \circ h, h^\star \gamma) = S(\phi, \gamma). \tag{2.4.23}$$

In the formalism of physics, we move the points in the domain by an infinitesimal diffeomorphism, that is, a vector field, and consider the variation

$$x^\alpha + \epsilon \delta x^\alpha \quad \text{or, in complex coordinates,} \quad z + \epsilon \delta z. \tag{2.4.24}$$

By (2.3.28), the conserved current is

$$j^z = -\left(\frac{\partial \phi}{\partial \bar{z}}\right)^2 \delta \bar{z}, \qquad j^{\bar{z}} = -\left(\frac{\partial \phi}{\partial z}\right)^2 \delta z, \tag{2.4.25}$$

and with $j_z = \gamma_{z\bar{z}} j^{\bar{z}}$ (cf. 1.1.2 and note that $\gamma_{zz} = 0$ by (2.4.14)), (2.3.31) becomes

$$0 = \frac{\partial}{\partial \bar{z}} j_z = \frac{\partial}{\partial \bar{z}} \left(\frac{\partial \phi}{\partial z} \right)^2 \delta z. \tag{2.4.26}$$

When we take holomorphic variations, $\frac{\partial}{\partial \bar{z}} \delta z = 0$, that is, respect the Riemann surface structure, this becomes (2.4.22), the holomorphicity of the energy–momentum tensor at a solution of the Euler–Lagrange equations, that is, (2.4.20).

We now wish to connect this discovery with 7 in Sect. 1.4.2. There, we had also found a holomorphic quadratic differential as a (co)tangent vector to the moduli space of Riemann surfaces. When we consider $S(\phi, \gamma)$ as a function of the metric γ, its derivative with respect to γ should be a tangent vector to the space of all metrics on our underlying surface. Here, we have been considering variations with respect to the inverse metric γ^{-1}, and thus, we obtain a cotangent instead of a tangent vector to the space of metrics. In 7 of Sect. 1.4.2, we have distinguished three types of variations of metrics, the ones through diffeomorphisms, the ones by conformal factors, and the residual ones that correspond to tangent directions of the Riemann moduli space. Now our functional $S(\phi, \gamma)$ is invariant under the first two types of variations: diffeomorphism invariance led to the holomorphicity (2.4.22), and conformal invariance made the energy–momentum tensor trace-free, (2.4.18). Therefore, it must correspond to a cotangent direction of the Riemann moduli space, and thus the agreement with the condition (1.4.20) is no coincidence.

2.4.2 The Nonlinear Sigma Model

In the nonlinear sigma model, the field ϕ takes its values in some Riemannian manifold N with metric g_{ij}, instead of in the real line $\mathbb{R}$. In the physics literature, one is usually interested in the case where N is the sphere S^n, that is, a homogeneous space for the Lie group $O(n+1)$ (one then speaks of the nonlinear $O(n+1)$ sigma model), or more generally, where N is the homogeneous space for some other compact Lie group. The case where N itself is a compact Lie group G leads to the Wess–Zumino–Witten model (see for instance [38, 73]). For the mathematical theory, however, one can consider an arbitrary Riemann manifold N, and this generality should make the structure more transparent. In fact, this will also be necessary for the applications to Morse theory presented below.

The action functional for the nonlinear sigma model is formally the same as (2.4.3),

$$S(\phi) = \frac{1}{2} \int_M \|d\phi\|^2 dvol(M), \tag{2.4.27}$$

where the norm of the differential is now given by

$$\|d\phi\|^2 = \gamma^{\alpha\beta}(x) g_{ij}(\phi(x)) \frac{\partial \phi^i(x)}{\partial x^\alpha} \frac{\partial \phi^j(x)}{\partial x^\beta}. \tag{2.4.28}$$

Expressed more abstractly, the differential of ϕ,

$$d\phi = \frac{\partial \phi^i}{\partial x^\alpha} dx^\alpha \otimes \frac{\partial}{\partial \phi^i}, \tag{2.4.29}$$

is now a section of the bundle $T^*M \otimes \phi^{-1}TN$ where the latter bundle is the pullback of the tangent bundle of the target N by the map ϕ. Since the bundle thus depends on the field ϕ, the situation is intrinsically nonlinear. In particular, the Euler–Lagrange equations are also nonlinear:

$$\tau^i(\phi) := \frac{1}{\sqrt{\gamma}} \frac{\partial}{\partial x^\alpha} \left(\sqrt{\gamma} \gamma^{\alpha\beta} \frac{\partial}{\partial x^\beta} \phi^i \right) + \gamma^{\alpha\beta}(x) \Gamma^i_{jk}(\phi(x)) \frac{\partial}{\partial x^\alpha} \phi^j \frac{\partial}{\partial x^\beta} \phi^k = 0, \tag{2.4.30}$$

with the Christoffel symbols as in (1.1.60). (The expression $\tau(\phi)$ is called the tension field of ϕ.) Whereas this nonlinearity makes the analysis more subtle and much harder, see [65], most of the formal aspects remain unchanged when compared with the linear version of Sect. 2.4. Solutions of (2.4.30) are called harmonic maps in the mathematical literature.

We are again interested in the situation when the underlying domain M is a Riemann surface. As in the linear case, the action (2.4.27) is conformally invariant, and so we can consider it either as a function $S(\phi, \gamma)$ of the domain metric γ or as a function $S(\phi, \Sigma)$, with the Riemann surface Σ considered as the equivalence class of conformal metrics that γ belongs to. Thus, conformal invariance is preserved in the nonlinear case, and so is, obviously, diffeomorphism invariance. Therefore, we again obtain a holomorphic energy–momentum tensor as before,

$$T_{zz} dz^2 = \left\langle \frac{\partial \phi}{\partial z}, \frac{\partial \phi}{\partial z} \right\rangle_N dz^2 = g_{ij}(\phi) \frac{\partial \phi^i}{\partial z} \frac{\partial \phi^j}{\partial z} dz^2 \tag{2.4.31}$$

where we use the scalar product defined by the metric of N. When one does the computation right, it is the same as in the linear case and therefore need not be repeated here.

For example, from (2.4.8), we also see that S is critical for variations of the metric γ when the energy–momentum tensor vanishes. According to (2.4.19), this means

$$\left\langle \frac{\partial \phi}{\partial z}, \frac{\partial \phi}{\partial z} \right\rangle_N = 0 \tag{2.4.32}$$

(note that we are not taking a Hermitian product here, and so this quantity can well be 0 without $\frac{\partial \phi}{\partial z}$ being 0 itself—when that happens, we say that $\frac{\partial \phi}{\partial z}$ is isotropic), or in real coordinates x, y with $z = x + iy$, from (2.4.9)

$$\left\langle \frac{\partial \phi}{\partial x}, \frac{\partial \phi}{\partial x} \right\rangle_N = \left\langle \frac{\partial \phi}{\partial y}, \frac{\partial \phi}{\partial y} \right\rangle_N, \qquad \left\langle \frac{\partial \phi}{\partial x}, \frac{\partial \phi}{\partial y} \right\rangle_N = 0. \tag{2.4.33}$$

When those relations hold, the map $\phi : \Sigma \to N$ is **conformal**. Since by a special case of the Riemann–Roch theorem, see Sect. 1.4.2, every holomorphic quadratic differential on the sphere S^2 vanishes, we conclude that on S^2, the energy–momentum tensor associated with a harmonic map automatically vanishes, and therefore, any harmonic map $\phi : S^2 \to N$ into any Riemann manifold N is conformal. For Riemann surfaces of genus > 0, this is not true.

2.4.3 The Supersymmetric Sigma Model

We now extend the sigma model to include supersymmetry, proceeding as in Sect. 2.2.5. We work with the Clifford algebra $Cl(2,0)$, which admits a real representation as explained in Sect. 1.3.2. In fact, this representation is a dimensional reduction of that of $Cl(2,1)$, and so the two-dimensional formalism to be developed here is a dimensional reduction of a three-dimensional one. As explained in [23], a three-dimensional space (with Minkowski signature) is the basic setting for $N = 1$ supersymmetry, but for our purposes, conformal invariance is a crucial underlying feature of our variational problems, and therefore, we continue to focus on the two-dimensional case and consider $(2|2)$ dimensions here (with Euclidean signature). We choose local even coordinates x^1, x^2 and odd ones θ^1, θ^2.

In order to conform to the conventions employed in [23], we use the following representation of $Cl(2,0)$:

$$e_1 \to \gamma^1 = \begin{pmatrix} -1 & 0 \\ 0 & 1 \end{pmatrix}, \qquad e_2 \to \gamma^2 = \begin{pmatrix} 0 & -1 \\ -1 & 0 \end{pmatrix} \tag{2.4.34}$$

which is different from the one described in Sect. 1.3.2 (but of course equivalent to it). We recall that the γ^μ satisfy

$$\{\gamma^\mu, \gamma^\nu\} = 2\delta^{\mu\nu}. \tag{2.4.35}$$

This is a real two-dimensional euclidean representation, and so we have real euclidean Majorana spinors satisfying

$$\bar{\psi} = \psi^\dagger = \psi^T = (\psi_1, \psi_2). \tag{2.4.36}$$

In particular, we could leave out the bars for complex conjugation (and we shall do so sometimes). Since these spinors are supposed to anticommute, we also have

$$\bar{\psi}\chi = -\bar{\chi}\psi, \qquad \bar{\psi}\gamma^\mu\chi = -\bar{\chi}\gamma^\mu\psi. \tag{2.4.37}$$

For a spinor field ψ, we then have the Dirac form

$$\bar{\psi}\mathbf{D}\psi = (\psi_1, \psi_2)\left(\gamma^1\begin{pmatrix} \frac{\partial\psi_1}{\partial x^1} \\ \frac{\partial\psi_2}{\partial x^1} \end{pmatrix} + \gamma^2\begin{pmatrix} \frac{\partial\psi_1}{\partial x^2} \\ \frac{\partial\psi_2}{\partial x^2} \end{pmatrix}\right)$$

$$= -\psi_1\frac{\partial\psi_1}{\partial x^1} + \psi_2\frac{\partial\psi_2}{\partial x^1} - \psi_1\frac{\partial\psi_2}{\partial x^2} - \psi_2\frac{\partial\psi_1}{\partial x^2}. \tag{2.4.38}$$

For later purposes, we observe that this can also be expressed in complex notation as

$$\bar{\psi} \not{D} \psi = -\left((\psi_1 + i\psi_2)\frac{\partial}{\partial z}(\psi_1 + i\psi_2) + (\psi_1 - i\psi_2)\frac{\partial}{\partial \bar{z}}(\psi_1 - i\psi_2)\right) \quad (2.4.39)$$

with $z = x^1 + ix^2$.

We define the vector fields

$$D_1 := \partial_{\theta^1} - \theta^1\partial_{x^1} - \theta^2\partial_{x^2}, \qquad (2.4.40)$$
$$D_2 := \partial_{\theta^2} - \theta^1\partial_{x^2} + \theta^2\partial_{x^1},$$

$$Q_1 := \partial_{\theta^1} + \theta^1\partial_{x^1} + \theta^2\partial_{x^2}, \qquad (2.4.41)$$
$$Q_2 := \partial_{\theta^2} + \theta^1\partial_{x^2} - \theta^2\partial_{x^1}.$$

They satisfy

$$[D_1, D_1] = -2\partial_{x^1}, \qquad [D_1, D_2] = -2\partial_{x^2}, \qquad [D_2, D_2] = 2\partial_{x^1}, \qquad (2.4.42)$$
$$[Q_1, Q_1] = 2\partial_{x^1}, \qquad [Q_1, Q_2] = 2\partial_{x^2}, \qquad [Q_2, Q_2] = -2\partial_{x^1}.$$

We are now ready to introduce the supersymmetric sigma model. We consider a superfield Y with expansion

$$Y = \phi(x) + \psi_\alpha(x)\theta^\alpha + F(x)\theta^1\theta^2 \qquad (2.4.43)$$

and the action

$$S_4 = \int \frac{1}{4}\epsilon^{\alpha\beta}\langle D_\alpha Y, D_\beta Y\rangle d^2x d\theta^2 d\theta^1 \qquad (2.4.44)$$

where $d\theta$ indicates that a Berezin integral has to be taken; namely, we recall from Sect. 1.5.2 that for an expression $Z = z + z_\alpha\theta^\alpha + z_{12}\theta^1\theta^2$,

$$\int Z d^2\theta = \int Z d\theta^2 d\theta^1 = z_{12}, \qquad (2.4.45)$$

that is, the θ-integration picks out the $\theta^1\theta^2$ term, see (1.5.27). Moreover, the antisymmetric ϵ-tensor is defined by

$$\epsilon^{12} = -\epsilon^{21} = 1. \qquad (2.4.46)$$

S_4 is the Wess–Zumino action (in flat Euclidean space). After expanding and carrying out the Berezin integral, this becomes

$$S_4 = \frac{1}{2}\int d^2x(\partial^\mu\phi^a\partial_\mu\phi_a + \bar{\psi}^a\gamma^\mu\partial_\mu\psi_a + F^a F_a). \qquad (2.4.47)$$

In complex notation, this looks as follows: We set

$$\psi_+ := \psi_1 - i\psi_2, \qquad \psi_- := \psi_1 + i\psi_2 \qquad (2.4.48)$$

and

$$\theta_+ := \theta_1 + i\theta_2, \qquad \theta_- := \theta_1 - i\theta_2 \tag{2.4.49}$$

and define the operators

$$D_+ := \partial_{\theta_+} + \theta_+ \partial_z, \qquad D_- := \partial_{\theta_-} + \theta_- \partial_{\bar{z}}. \tag{2.4.50}$$

With this notation, (2.4.43) becomes

$$Y = \phi + \frac{1}{2}(\psi_+ \theta_+ + \psi_- \theta_-) + \frac{i}{2} F \theta_+ \theta_-. \tag{2.4.51}$$

We then have

$$S_4 = \int \frac{1}{2} D_- Y D_+ Y d^2 x d\theta_- d\theta_+$$

$$= \int \frac{1}{2} \left(4\partial_z \phi^a \partial_{\bar{z}} \phi_a - \psi_+ \frac{\partial}{\partial \bar{z}} \psi_+ - \psi_- \frac{\partial}{\partial z} \psi_- + F^a F_a \right) d^2 x. \tag{2.4.52}$$

S_4 is invariant under the supersymmetry transformations

$$\delta \phi^a = \bar{\varepsilon} \psi^a, \tag{2.4.53}$$

$$\delta \psi^a = \gamma^\mu \varepsilon \partial_\mu \phi^a - \bar{\varepsilon} F^a, \tag{2.4.54}$$

$$\delta F^a = \varepsilon \gamma^\mu \partial_\mu \psi^a. \tag{2.4.55}$$

Indeed, the variation is

$$\delta S_4 = \frac{1}{2} \int dx^2 \left(2\bar{\varepsilon} \partial^\mu \phi^a \partial_\mu \psi_a + \bar{\varepsilon} \gamma^\nu \partial_\nu \phi^a \gamma^\mu \partial_\mu \psi_a + \bar{\varepsilon} F^a \gamma^\mu \partial_\mu \psi_a \right.$$

$$\left. + \overline{\psi}^a \gamma^\mu \partial_\mu (\gamma^\nu \varepsilon \partial_\nu \phi_a + \varepsilon F_a) - 2\bar{\varepsilon} \gamma^\mu \partial_\mu \psi^a \, F_a + \partial_\mu (\bar{\varepsilon}) \psi^a \partial^\mu \phi_a \right), \tag{2.4.56}$$

which vanishes after integration by parts and using (2.4.35) and (2.4.37), when we assume that $\bar{\varepsilon}$ is constant. Locally, the latter can be assumed, and we do so for the moment, but later on, in Sect. 2.4.7, we shall return to the global issue.

The Euler–Lagrange equations for S_4 are

$$\epsilon^{\alpha\beta} D_\alpha D_\beta Y = 0, \tag{2.4.57}$$

or in components,

$$\Delta \phi = 0, \tag{2.4.58}$$

$$\gamma^\mu \partial_\mu \psi = 0, \tag{2.4.59}$$

$$F = 0. \tag{2.4.60}$$

In complex notation, these equations become

$$D_- D_+ Y = 0, \tag{2.4.61}$$

or in components

$$\partial_z \partial_{\bar{z}} \phi = 0, \tag{2.4.62}$$

$$\partial_{\bar{z}} \psi_+ = 0, \qquad \partial_z \psi_- = 0, \tag{2.4.63}$$

$$F = 0. \tag{2.4.64}$$

Again, F is a nonpropagating, auxiliary field that is only introduced to close the supersymmetry algebra off-shell. On-shell, (2.4.53) and (2.4.54) become

$$\delta \phi^a = \bar{\varepsilon} \psi^a, \tag{2.4.65}$$

$$\delta \psi^a = \gamma^\mu \varepsilon \partial_\mu \phi^a. \tag{2.4.66}$$

(Note that (2.4.59) implies that $\delta F^a = 0$ on-shell, i.e., the term that obstructs the closing of the algebra is proportional to one to the equations of motion.)

We now turn to the supersymmetric *nonlinear* sigma model. In fact, the formalism remains the same; we just need to expand its interpretation.

Thus, we consider a map

$$Y : M \to N \tag{2.4.67}$$

from a $(2|2)$-dimensional supermanifold to some Riemannian manifold N. We expand Y as before:

$$Y = \phi(x) + \psi_\alpha(x) \theta^\alpha + F(x) \theta^1 \theta^2. \tag{2.4.68}$$

ϕ can be considered to be an ordinary map into N, whereas the odd part ψ represents an (odd) section of the pull-back tangent bundle $\phi^\star TN$.[17] $\langle . , . \rangle$ now denotes a Riemannian metric on the target space; we shall also write $\|v\|^2 := \langle v, v \rangle$ below.

Finally, F is an auxiliary field as before. This time, the algebraic equation for F among the Euler–Lagrange equations is

$$-4 g_{ij}(\phi) F^i + 2 g_{ij,k} \overline{\psi}^k \psi^i - g_{ki,j} \overline{\psi}^k \psi^i = 0,$$

i.e.,

$$F^i = \Gamma^i_{kj} \overline{\psi}^k \psi^j, \tag{2.4.69}$$

so that F can again be eliminated. In particular, when we use Riemann normal coordinates at the point under consideration, F vanishes.

---

[17]Thus, w.r.t. coordinate changes on the target N, ψ transforms as a vector, whereas on the domain M, it transforms as a spinor. In particular, the setting here is different from the one above in Sect. 1.5.3 for maps between super Riemann surfaces, where the odd field has to transform as a spinor on both domain and target.

In local coordinates, after carrying out the θ-integral, the Lagrangian density becomes

$$\frac{1}{2}\|d\phi\|^2 + \frac{1}{2}\langle \psi, \not{D}\psi \rangle - \frac{1}{12}\epsilon^{\alpha\beta}\epsilon^{\gamma\delta}\langle \psi_\alpha, R(\psi_\beta, \psi_\gamma)\psi_\delta \rangle. \tag{2.4.70}$$

(We get the R-term (the curvature of the target N) after elimination of $\frac{1}{2}\|F\|^2$.) This comes about as follows: According to the rule for the Berezin integral, we need to identify the $\theta^1\theta^2$-term in (2.4.44). For that purpose, we recall that a function of a superfield Y has to be expanded by Taylor's formula as explained in Sect. 1.5.2, see (1.5.20), (1.5.21). In particular, (2.4.44) contains the metric tensor $\langle .,. \rangle$ of the target N. In local coordinates, we have a tensor $g_{ij}(Y)$ whose expansion contains second derivatives $g_{ij,kl}(\phi)$ multiplied with $\theta^1\psi_1\theta^2\psi_2$, which gives the curvature term in (2.4.70). Terms with first derivatives of g_{ij} do not carry an invariant meaning and become 0 in suitable coordinates (Riemann normal coordinates) at the point in N under consideration. In particular, the curvature tensor R has to be evaluated at the point $\phi(x) \in N$. Similarly, the Dirac operator $\not{D}$ contains a covariant derivative at the tangent space $T_{\phi(x)}N$.

We now list the important results for the nonlinear supersymmetric sigma model (for detailed computations see [17]): The Euler–Lagrange equations for the nonlinear supersymmetric sigma model are

$$\tau^m(\phi) - \frac{1}{2}R^m{}_{lij}\overline{\psi}^i(\nabla\phi^l \cdot \psi^j) + \frac{1}{12}g^{mp}R_{ikjl;p}(\overline{\psi}^i\psi^j)(\overline{\psi}^k\psi^l) = 0, \tag{2.4.71}$$

$$\not{D}\psi^m - \frac{1}{3}R^m_{jkl}\psi^k\overline{\psi}^j\psi^l = 0. \tag{2.4.72}$$

The first equation generalizes (2.4.30).

The functional S is invariant under the supersymmetry transformation

$$\begin{cases} \delta\phi^i = \varepsilon\psi^i, \\ \delta\psi^i = \gamma^\alpha\partial_\alpha\phi^i\varepsilon - \Gamma^i_{jk}(\varepsilon\psi^j)\psi^k. \end{cases} \tag{2.4.73}$$

As in (2.4.55), we recognize the F-term, see (2.4.69).

The supercurrent

$$J^\alpha := 2g_{ij}\partial_\beta\phi^i\gamma^\beta\gamma^\alpha\psi^j, \quad \alpha = 1, 2 \tag{2.4.74}$$

is conserved (on-shell), i.e.,

$$D_\alpha J^\alpha \equiv 0. \tag{2.4.75}$$

Again, we wish to consider an interaction Lagrangian with a superpotential W

$$S_{int} = \int W(Y(x,\vartheta))d^2x d^2\theta. \tag{2.4.76}$$

A simple and standard choice is

$$W(Y) = -\frac{1}{3}kY^3 - \lambda Y, \qquad (2.4.77)$$

with parameters k, λ. The coefficient of $\theta^1 \theta^2$ in the expansion of W is

$$-\lambda F - kF\phi^2 - k\overline{\psi}\psi\phi.$$

We first consider the linear sigma model, that is, we set the curvature tensor $R = 0$. In the Lagrangian $S_4 + S_{int}$, we then have the F terms

$$\frac{1}{2}F^2 + \lambda F + kF\phi^2,$$

leading to the algebraic Euler–Lagrange equation

$$F = -\lambda - k\phi^2.$$

Utilizing this equation, the Lagrangian becomes

$$S_4 + S_{int} = \int d^2x \left(\frac{1}{2}\partial_\mu\phi\partial^\mu\phi + \frac{1}{2}\overline{\psi}\gamma^\mu\partial_\mu\psi - \frac{1}{2}(\lambda + k\phi^2)^2 - k\overline{\psi}\psi\phi \right).$$

A more general interaction term is of the form

$$S_{int} = \frac{1}{2} \int h(Y)d^2x d^2\theta = \int \left(-\frac{1}{2}g^{ij}(\phi)\frac{\partial h}{\partial\phi^i}\frac{\partial h}{\partial\phi^j} - \frac{1}{2}\frac{\partial^2 h}{\partial\phi^i\partial\phi^j}\overline{\psi}^i\psi^j \right)d^2x$$

$$(2.4.78)$$

where the first term in the integrand arises from eliminating the auxiliary field F in the combined Lagrangian, in the same manner as before.

2.4.4 Boundary Conditions

We start again with the bosonic field ϕ that takes its values in some d-dimensional Riemannian manifold N. We now assume that ϕ is defined on some Riemann surface with boundary. The surface will again be denoted by Σ, and we assume for the moment that its boundary is a smooth curve γ, or a collection of such curves. Boundary conditions for ϕ on $\gamma = \partial\Sigma$ are given by specifying a smooth submanifold B of dimension p of N. In the physics literature [86, 87], this would be called a **brane** or a **D-brane**, with D standing for Dirichlet boundary conditions.[18] Locally, we can choose coordinates on N so that B is given by $x^{p+1} = \cdots = x^d = 0$.

---

[18]In fact, the physics convention is to consider $d - 1$ spatial and one temporal dimensions. A p-brane would then result from fixing p of the $d - 1$ spatial dimensions.

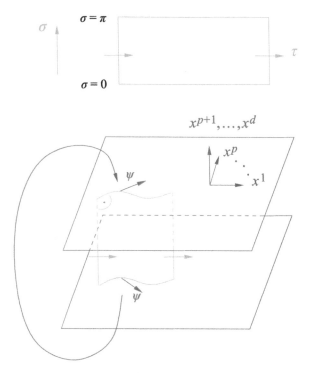

Fig. 2.1 The boundary conditions for bosonic and fermionic fields defined by a D-brane

The boundary conditions are illustrated in Fig. 2.1 and will now be described in formulae.

The first part of the boundary conditions requires that the boundary curve γ be mapped to B. Locally, this therefore means

$$\phi^{p+1} = \cdots = \phi^d = 0 \quad \text{on } \gamma. \tag{2.4.79}$$

This is, of course, a Dirichlet boundary condition for the components $\phi^{p+1}, \ldots, \phi^d$. More generally, the tangential derivative $\frac{\partial}{\partial \tau}$ of these components has to vanish on γ,

$$\frac{\partial \phi^{p+1}}{\partial \tau} = \cdots = \frac{\partial \phi^d}{\partial \tau} = 0 \quad \text{on } \gamma. \tag{2.4.80}$$

The remaining components then should satisfy a Neumann boundary condition, that is, with $\frac{\partial}{\partial \nu}$ denoting the normal derivative on γ,

$$\frac{\partial \phi^0}{\partial \nu} = \cdots = \frac{\partial \phi^p}{\partial \nu} = 0 \quad \text{on } \gamma. \tag{2.4.81}$$

In order to state the boundary condition for the fermionic field ψ, we use the notation of (2.4.48). We first assume that at the point under consideration, the metric of N is given in normal coordinates, that is, $g_{ij} = \delta_{ij}$. Then the general boundary condition for ψ is in these coordinates:

$$\psi_-^j = \pm \psi_+^j. \tag{2.4.82}$$

A choice of sign in (2.4.82) can be motivated by the following consideration. As our domain, we consider a strip

$$\{\sigma \in [0, 2\pi]\} \times \{\tau \in [0, T]\}, \tag{2.4.83}$$

and we wish to fix boundary conditions for $\sigma = 0, 2\pi$ (since the τ-direction is interpreted as a temporal direction, there might be initial conditions prescribed at $\tau = 0$ and final ones at $\tau = T$, but this is not our concern here). We assume that we have a mapping ϕ defined on this strip, and a vector ψ along ϕ. We look at the simplest situation, where N is Euclidean space $\mathbb{R}^d$, and the brane receiving $\sigma = 0$ is the hyperplane $x^{d+1} = \ldots x^d = 0$. For $\sigma = 2\pi$, we prescribe another brane that is parallel to the first one, say $x^{d+1} = \ldots x^{d-1} = 0, x^d = R$. When we then periodically identify these two branes, that is, dividing $\mathbb{R}^d$ by the translations by R in the x^d-direction, in order to get the fields ψ^j to match on the boundary, we need to require

$$\psi^k(2\pi, \tau) = \psi^k(0, \tau) \quad \text{for } k = 1, \ldots, p, \tag{2.4.84}$$

and

$$\psi^\ell(2\pi, \tau) = -\psi^\ell(0, \tau) \quad \text{for } \ell = p+1, \ldots, d. \tag{2.4.85}$$

When we then put $\psi_+ = \psi$ and define $\psi_-(\sigma, \tau) = \psi_+(2\pi - \sigma, \tau)$ (in which case the field ψ on the range $\sigma \in [0, 2\pi]$ is obtained from the two fields $\psi_\pm$ on half the range, $\sigma \in [0, \pi)$), we then obtain from (2.4.84)

$$\psi_-^j = \psi_+^j \quad \text{for } j = 1, \ldots, p, \quad \text{and} \quad \psi_-^k = -\psi_+^k \quad \text{for } k = p+1, \ldots, d. \tag{2.4.86}$$

Thus, the plus sign in (2.4.82) corresponds to Neumann boundary conditions for the corresponding components of ϕ, and the minus sign to Dirichlet conditions. See also the discussion in Sect. 2.6.3, around (2.6.47).

In general coordinates, the boundary conditions for the ψ-field can be written as

$$\psi_-^j = D_i^j \psi_+^i, \tag{2.4.87}$$

with the tensor D_i^j satisfying

$$D_i^j D_k^i = \delta_k^j \tag{2.4.88}$$

and

$$g_{ij} = D_i^k D_j^\ell g_{k\ell}. \tag{2.4.89}$$

Thus

$$D_{ij} = g_{ik} D_j^k \tag{2.4.90}$$

is symmetric. Again, there is a sign to be determined, according to Neumann or Dirichlet type boundary conditions for ϕ.

The preceding boundary conditions arise as follows. When we consider variations $\delta\phi, \delta\psi$ of the fields in S_4 (neglecting the F-field as this vanishes on-shell anyway), we get a corresponding boundary term in the induced variation δS_4. Employing the version (2.4.52), this boundary term is given by

$$\frac{1}{2}\int g_{jk}\left(\delta\phi^j\frac{\partial\phi^k}{\partial\nu} + i(\delta\psi_+^j\psi_+^k - \delta\psi_-^j\psi_-^k) + i\delta\phi^j\Gamma_{\ell m}^k(\psi_-^\ell\psi_-^m - \psi_+^\ell\psi_+^m)\right)d\xi$$

$$(2.4.91)$$

where ν as before is the outer normal direction at γ and ξ is a coordinate on γ. These boundary terms then vanish if

1. either

$$\delta\phi^j = 0 \quad \text{(Dirichlet)} \qquad (2.4.92)$$

or

$$\frac{\partial\phi^k}{\partial\nu} = 0 \quad \text{(Neumann)} \qquad (2.4.93)$$

2. and

$$\psi_-^j = D_i^j\psi_+^i, \qquad (2.4.94)$$

that is, (2.4.87) holds, with the conditions (2.4.88) and (2.4.89).

In order to keep the theory supersymmetric, the brane B should be totally geodesic, that is, every shortest geodesic in N connecting two points in B should already be contained in B, see [1].

2.4.5 Supersymmetry Breaking

The Hilbert space of a quantum field theory can be decomposed as

$$\mathcal{H} = \mathcal{H}^+ \oplus \mathcal{H}^-$$

with $\mathcal{H}^+(\mathcal{H}^-)$ being the space of "bosonic" ("fermionic") states. The theory is supersymmetric if there are (Hermitian) supersymmetry operators

$$Q_i : \mathcal{H} \to \mathcal{H}, \quad i = 1, \dots, N$$

with

$$Q_i(\mathcal{H}^\pm) = \mathcal{H}^\mp. \qquad (2.4.95)$$

Witten [104] introduced the operator $(-1)^F$ satisfying

$$(-1)^F\chi = \pm\chi \quad \text{for } \chi \in \mathcal{H}^\pm. \qquad (2.4.96)$$

The supersymmetry operators Q_i then anticommute with $(-1)^F$

$$(-1)^F Q_i + Q_i (-1)^F = 0 \quad (i = 1, \dots, N).$$ (2.4.97)

The Q_i must commute with the Hamiltonian H that generates the time translations, i.e.,

$$Q_i H - H Q_i = 0 \quad (i = 1, \dots, N).$$ (2.4.98)

The Q_i are then determined if we require additionally

$$Q_i^2 = H, \qquad Q_i Q_j + Q_j Q_i = 0 \quad \text{for } i \neq j.$$ (2.4.99)

As a square of Hermitian operators, H is positive semidefinite.

Let $|b\rangle \in \mathcal{H}^+$ satisfy

$$H|b\rangle = E|b\rangle,$$

i.e., $|b\rangle$ is an eigenvector H with eigenvalue E (≥ 0 as H is positive semidefinite). We consider one of the Q_i, which we simply write as Q for the moment.

We can write

$$Q|b\rangle = \sqrt{E}|f\rangle$$

and get

$$Q|f\rangle = \frac{1}{\sqrt{E}} Q^2 |b\rangle = \frac{1}{\sqrt{E}} H|b\rangle = \sqrt{E}|b\rangle.$$

Thus, if $E \neq 0$, the bosonic and fermionic eigenstates with eigenvalue E are paired in an irreducible multiplet of the supersymmetry algebra. This need not be so any longer if $E = 0$. Since $H = Q^2$, if

$$H|b_0\rangle = 0 \quad \text{for } |b_0\rangle \in \mathcal{H}^+$$

we have

$$0 = \langle b_0|H|b_0\rangle = \| Q|b_0\rangle \|^2 \quad \text{(since } Q \text{ is hermitian)},$$

hence

$$Q|b_0\rangle = 0,$$

and similarly $H|f_0\rangle = 0$ for $|f_0\rangle \in \mathcal{H}^-$ implies

$$Q|f_0\rangle = 0.$$

Thus, the zero eigenvectors of H are supersymmetric, that is, invariant under the supersymmetry operator.

Consequently, for positive energy E, the number of bosonic eigenvectors equals the number of fermionic ones, but this need not be so for zero energy.

Let

$$\nu(0) := \#\text{bosonic} - \#\text{ fermionic zero eigenvectors.}$$

Regularizing the trace of $(-1)^F$, we then have

$$\text{Tr}(-1)^F = \nu(0). \tag{2.4.100}$$

If $\nu(0) \neq 0$, there must exist at least one—bosonic or fermionic—state with zero energy. Since 0 is the smallest possible value of the energy, such a state furnishes a vacuum that is supersymmetric. If there does not exist a supersymmetric vacuum, i.e., if the smallest eigenvalue of H is positive, one says that supersymmetry is spontaneously broken.

We first consider the supersymmetric point particle with two odd variables with the total Lagrangian (see (2.2.86), (2.2.99), (2.2.101))

$$L_3 + L_{int} = \int dt \left(\frac{1}{2}\dot{\phi}\dot{\phi} + \frac{1}{2}\psi_\alpha \dot{\psi}_\alpha - \frac{1}{2}\left(\frac{dw(\phi)}{d\phi}\right)^2 - \frac{d^2w}{d\phi^2}\psi_1\psi_2 \right). \tag{2.4.101}$$

The Hamiltonian is

$$H = \frac{1}{2}p^2 + \frac{1}{2}\left(\frac{dw}{d\phi}\right)^2 + \frac{d^2w}{d\phi^2}\psi_1\psi_2. \tag{2.4.102}$$

As before, see (2.4.77), we choose

$$w(\phi) = -\frac{1}{3}k\phi^3 - \lambda\phi. \tag{2.4.103}$$

We obtain

$$H = \frac{1}{2}p^2 + \frac{1}{2}(k\phi^2 + \lambda)^2 + 2k\phi\psi_1\psi_2. \tag{2.4.104}$$

ψ_1 and ψ_2 are Grassmann valued and odd, and so

$$[\psi_\alpha, \psi_\beta] = \psi_\alpha\psi_\beta + \psi_\beta\psi_\alpha = 0 \quad \text{for } \alpha, \beta = 1, 2. \tag{2.4.105}$$

After quantization, we get, in place of (2.4.105),

$$[\psi_\alpha, \psi_\beta] = \hbar\delta_{\alpha\beta}, \tag{2.4.106}$$

that is, the Grassmann variables become Clifford algebra valued.

We may thus represent the ψ_α by Pauli matrices

$$\psi_\alpha = \sqrt{\frac{1}{2}\hbar}\,\sigma_\alpha \tag{2.4.107}$$

and get

$$H = \frac{1}{2}p^2 + \frac{1}{2}(k\phi^2 + \lambda)^2 + \hbar k\sigma_3\phi. \tag{2.4.108}$$

At the so-called tree level (that is, keeping only the zeroth-order terms (those that are not proportional to $\hbar^n, n > 0$)—the higher order contain corrections of the tree level), the ground state energy is determined by the potential $(k\phi^2 + \lambda)^2$. Hence, supersymmetry is spontaneously broken if $\frac{\lambda}{k} > 0$.

The supersymmetry generators here are

$$Q_1 = \frac{1}{\sqrt{2}}(\sigma_1 p + \sigma_2(k\phi^2 + \lambda)), \qquad Q_2 = \frac{1}{\sqrt{2}}(\sigma_2 p - \sigma_1(k\phi^2 + \lambda)). \quad (2.4.109)$$

For the sequel, it will be convenient to switch to the operators

$$Q_\pm = \frac{1}{\sqrt{2}}(Q_1 \pm i Q_2). \qquad (2.4.110)$$

Then

$$Q_\pm^2 = 0, \qquad [Q_+, Q_-] = 2H. \qquad (2.4.111)$$

In a cohomological interpretation, we call a state $|s\rangle \in \mathcal{H}$ with

$$Q_+|s\rangle = 0$$

closed, one that can be written as

$$|s\rangle = Q_+|t\rangle \quad \text{for some } |t\rangle \in \mathcal{H}$$

exact. Since $Q_+^2 = 0$, exact states are closed. Conversely, if $|s_E\rangle$ is a closed eigenvector of H with eigenvalue $E \neq 0$, i.e.,

$$H|s_E\rangle = E|s_E\rangle,$$

then $|t_E\rangle := \frac{1}{E}Q_-|s_E\rangle$ satisfies

$$Q_+|t_E\rangle = \frac{1}{E}Q_+Q_-|s_E\rangle = \frac{1}{E}[Q_+, Q_-]|s_E\rangle \quad (Q_-Q_+|s_E\rangle = 0 \text{ as } s_E \text{ is closed})$$

$$= \frac{1}{E}H|s_E\rangle = |s_E\rangle,$$

and hence $|s_E\rangle$ is exact.

If $E = 0$ and if we had again $|s_0\rangle = Q_+|t_0\rangle$, then also

$$H|t_0\rangle = 0 \quad \text{as } [Q_+, H] = 0.$$

However, by (2.4.99), this implies $Q_1|t_0\rangle = Q_2|t_0\rangle = 0$ as above, hence also

$$|s_0\rangle = Q_+|t_0\rangle = 0.$$

Thus, the nonvanishing eigenstates for $E = 0$ are precisely the non-exact closed states.

2.4.6 The Supersymmetric Nonlinear Sigma Model
and Morse Theory

We return to the supersymmetric nonlinear sigma model. We let N be a compact Riemannian manifold. We assume that the so-called world sheet, the two-dimensional domain on which the fields are defined, is of the form

$$\{(t, x) : t \in \mathbb{R}, x \in S^1\},$$

i.e., a cylinder, whose circumference we assume to have length L. We also assume that the fields ϕ^i and ψ^i are independent of x. We then get

$$S_5 = \frac{1}{2} L \int dt \left(g_{ij}(\phi) \frac{\partial \phi^i}{\partial t} \frac{\partial \phi^j}{\partial t} + g_{ij}(\phi) \overline{\psi}^i \gamma^0 \partial_t \psi^j + \frac{1}{6} R_{ijkl} \overline{\psi}^i \psi^k \overline{\psi}^j \psi^l \right).$$

$$(2.4.112)$$

After quantization, the spinors ψ^i and their Hermitian conjugates become Clifford algebra valued, i.e.,

$$[\psi^i, \psi^j] = 0 = [\psi^{i*}, \psi^{j*}], \qquad [\psi^i, \psi^{j*}] = g_{ij}(\phi).$$

Also, after quantization, supersymmetry is generated by the charges

$$Q_+ = i \psi^{i*} p_i = \psi^{i*} D_{\phi^i}, \qquad Q_- = -i \psi^i p_i = -\psi^i D_{\phi^i},$$

with D_{ϕ^i} being a covariant derivative, the momentum conjugate to ϕ^i.

We now recall from Sect. 1.3.2 that we have a representation of the Clifford algebra on the space of spinors given by

$$\psi^{j*} \sim \varepsilon(dx^j) \quad (\varepsilon(dx^j) \text{ operates as the exterior product}$$

with the differential form dx^j),

$$\psi^i \sim i(dx^i) \quad (i(dx^i) \text{ operates as interior contraction with } dx^i).$$

(This representation is obtained from the one in Sect. 1.3.2 by setting the imaginary parts of the differential forms to 0.)

Thus, ψ^{j*} corresponds to a differential form, ψ^i to a vector field on N (here $x^1, x^2, \ldots$, are local coordinates on N; one should write $\phi^1, \phi^2, \ldots$ in place of $x^1, x^2, \ldots$, but expressions like $d\phi^i$ look a bit awkward).

Moreover, Q_+ then corresponds to the exterior derivative d, Q_- to its adjoint d^*, and the Hamiltonian is

$$H = Q_+ Q_- + Q_- Q_+ = dd^* + d^*d, \qquad (2.4.113)$$

the Hodge Laplacian (1.1.109). With $Q_1 := d + d^*$, $Q_2 := i(d - d^*)$, we also have

$$H = Q_1^2 = Q_2^2. \qquad (2.4.114)$$

On the other hand, we had interpreted ψ^{j^*} as a fermionic creation operator, ψ^i as a fermionic annihilation operator. The states that are annihilated by all ψ^i, i.e., the states with no fermions, are then identified with the functions $f(x)$ on N. Operating on such a state with a ψ^{i^*}, we obtain a state with one fermion, or in the de Rham picture (see the discussion at the end of Sect. 1.1.3), a one-form on N. States with two fermions must be antisymmetric in the fermionic indices, because of the fermion statistics, and can be considered as two-forms.

Thus, we obtain the de Rham complex, with the Hodge Laplacian. The dimension of the space of zero states of this Laplacian, i.e., of harmonic q-forms, is the Betti number b_q.

Equating the two pictures gives Witten's result [104]

$$\mathrm{Tr}(-1)^F = \sum_q (-1)^q b_q(N).$$

We now add our self-interaction term L_{int} with Morse function sh (s here is a parameter) to the Lagrangian S_5. (A smooth (twice continuously differentiable) function h is called a Morse function if at all its critical points the Hessian, that is, the matrix of its second derivatives, is nondegenerate, that is, does not have 0 as an eigenvalue.) This changes d, d^* to

$$d_s = e^{-hs} d e^{hs}, \qquad d_s^* = e^{hs} d^* e^{-hs}. \tag{2.4.115}$$

We have $d_s^2 = 0 = d_s^{*2}$, and we get

$$Q_{1,s} = d_s + d_s^*, \qquad Q_{2,s} = i(d_s - d_s^*). \tag{2.4.116}$$

Moreover,

$$\begin{aligned} H_s &= Q_{1,s}^2 = Q_{2,s}^2 \\ &= d_s d_s^* + d_s^* d_s \\ &= dd^* + d^* d + s^2 g^{ij} \frac{\partial h}{\partial x^i} \frac{\partial h}{\partial x^j} + s \frac{\partial^2 h}{\partial x^i \partial x^j} [\varepsilon(dx^i), i(dx^j)]. \end{aligned} \tag{2.4.117}$$

$s^2 g^{ij} \frac{\partial h}{\partial x^i} \frac{\partial h}{\partial x^j}$ is the potential energy, and it becomes very large for large s, except in the vicinity of the critical points of h. Therefore, the eigenfunctions of H_s concentrate near the critical points of h for large s, and asymptotic expansions in powers of $\frac{1}{s}$ for the eigenvalues depend only on local data near the critical points. This is the starting point of Witten's approach to Morse theory [105], which we shall now discuss.

As mentioned, we assume that h is a Morse function. We let $q_1, q_2, \ldots, q_m$ be the critical points of h. By the Morse lemma (see e.g. [65], p. 311), each critical point q_v has a neighborhood U_v with the property that in suitable local coordinates $x = x_v = (x_v^1, \ldots, x_v^n)$ with $x_v(q_v) = 0$,

$$h(p) - h(q_v) = \frac{1}{2} \sum_{k=1}^n \mu_{v,k} x_v^k(p)^2 \tag{2.4.118}$$

with

$$D^2h(q_v) = \text{diag}(\mu_{v,1}, \ldots, \mu_{v,n}) \qquad (2.4.119)$$

(i.e., the Hessian of h at q_v is diagonalized, and the diagonal elements $\mu_{v,1}, \ldots, \mu_{v,n}$ are nonzero as h is assumed to be a Morse function).

Also, on U_v we choose a flat Riemannian metric g_v for which the $\frac{\partial}{\partial x_v^j}, j = 1, \ldots, n$, are orthonormal.

Of course, we may assume that the $U_v, v = 1, \ldots, m$, are pairwise disjoint, and moreover that their closures are contained in pairwise disjoint open sets V_v.

$$V_0 := N \setminus \bigcup_{v=1}^{m} \overline{U_v}, \quad V_1, \ldots, V_m$$

is then an open covering of N, and we may find a subordinate partition of unity $\{\eta_v\}_{v=0}^{m}$, that is, functions $\eta_v : N \to \mathbb{R}$ satisfying

$$0 \le \eta_v \le 1, \qquad \sum_{v=0}^{m} \eta_v = 1, \qquad \text{supp}\,\eta_v \subset V_v,$$

with $\eta_v = 1$ on U_v.

We choose any metric g_0 on V_0 and put

$$g := \sum_{v=0}^{m} \eta_v g_v.$$

g is then a Riemannian metric on N. Since neither the Betti numbers of N nor the critical points of h or their Morse indices depend on the choice of a Riemannian metric on N, we may work with the metric g in the sequel. In this metric, we have on $U_v (v = 1, \ldots, m)$

$$H_s = \sum_{j=1}^{n} \left(-\left(\frac{\partial}{\partial x^j} \right)^2 + s^2 \mu_{v,j}^2 x^{j^2} + \frac{1}{s} \mu_{v,j} [\varepsilon(dx^j), i(dx^j)] \right). \qquad (2.4.120)$$

In particular, H_s is an operator with separated variables on U_v. In fact, we have

$$H_s = \sum_{j=1}^{n} \left(\Omega_s^{j,v} + \frac{1}{s} \mu_{v,j} K^j \right) \qquad (2.4.121)$$

with

$$\Omega_s^{j,v} := -\left(\frac{\partial}{\partial x^j} \right)^2 + s^2 \mu_{v,j}^2 x^{j^2}, \qquad (2.4.122)$$

$$K^j := [\varepsilon(dx^j), i(dx^j)]. \qquad (2.4.123)$$

The operators $\Omega_s = \Omega_s^{j,\nu}$ are just Hamiltonians for harmonic oscillators, and they have eigenvalues

$$s|\mu_{\nu,j}|(1+2N) \quad (N = 0, 1, 2, \ldots) \tag{2.4.124}$$

with eigenfunctions

$$\phi_N(x) = P_N\left(\sqrt{s|\mu_{\nu,j}|}x\right)e^{-\frac{s}{2}|\mu_{\nu,j}|x^2}, \tag{2.4.125}$$

where the P_N are Hermite polynomials. In particular, for large s, the ϕ_N rapidly decay away from $x = 0$, i.e., away from the critical point q_ν of h. Moreover, we have

$$K^j dx^{\alpha_1} \wedge \cdots \wedge dx^{\alpha_r} = \varepsilon_j^{\boldsymbol{\alpha}}(dx^{\alpha_1} \wedge \cdots \wedge dx^{\alpha_r}), \tag{2.4.126}$$

with

$$\varepsilon_j^{\boldsymbol{\alpha}} = \begin{cases} 1, & \text{if } j \in \boldsymbol{\alpha} = (\alpha_1 \ldots \alpha_r), \\ -1, & \text{otherwise} \end{cases} \tag{2.4.127}$$

i.e., K^j has eigenvalues ± 1. H_s thus is a self-adjoint operator with eigenvalues

$$s\sum_{j=1}^n((1+2N_{\nu,j})|\mu_{\nu,j}| + \varepsilon_{\nu,j}\mu_{\nu,j}), \tag{2.4.128}$$

$\varepsilon_{\nu,j} = \pm 1$, $N_{\nu,j} = 0, 1, 2, \ldots$, and orthonormal eigenvectors

$$\phi_{N_\nu,\alpha_\nu}^s = s^{\frac{n}{4}} \prod_{j=1}^n P_{N_{\nu,j}}\left(\sqrt{s|\mu_{\nu,j}|}x^j\right)$$

$$\times \left[\exp\left(-\frac{s}{2}\sum_{j=1}^n|\mu_{\nu,j}|x^{j^2}\right)dx^{\alpha_{\nu,1}} \wedge \cdots \wedge dx^{\alpha_{\nu,r}}\right]$$

(with $\boldsymbol{\alpha}_\nu = (\alpha_{\nu,1}, \ldots, \alpha_{\nu,r})$). $\tag{2.4.129}$

In order for an eigenvalue to vanish, we necessarily have $N_{\nu,j} = 0$ for all j, and moreover $\varepsilon_{\nu,j}$ and $\mu_{\nu,j}$ have opposite signs. Thus, if p_ν has Morse index p, i.e., precisely p of the $\mu_{\nu,j}$ are negative, then p out of the $\varepsilon_{\nu,j}$ must be positive, which means that the corresponding eigenvector is a p-form, as can be seen from (2.4.126) and (2.4.127). Thus, if a_ν is a critical point of Morse index p, it has a one-dimensional contribution to the nullspace of H_s operating on p-forms, while for different Morse index, there is no contribution.

Now this has been a local consideration, and a nulleigenvector on U_ν need not extend to a nulleigenvector on all of N. However, a perturbation argument (see, e.g., [57] or the monograph [15] for details) shows that the other eigenvalues of H_s on $\Lambda^p(N)$ diverge as s tends to ∞, while the global nulleigenvectors concentrate at the critical points and therefore lead to local nulleigenvectors as considered above. We conclude the basic theorem of Morse:

Theorem 2.1

$$m_p \geq b_p, \tag{2.4.130}$$

where m_p is the number of critical points of h of Morse index p and b_p is the pth Betti number of N.

(b_p is the dimension of the kernel of the Hodge Laplacian $dd^* + d^*d$ on $\Omega^p(N)$, and one easily sees that the dimension of the kernel of the perturbed Laplacian $H_s = d_s d_s^* + d_s^* d_s$ is the same for all s.)

Of course, this is an asymptotic argument, for $s \to \infty$, and we only get expansions of the eigenvectors, in contrast to the original case $s = 0$ where we could identify them with harmonic forms. However, here already the classical, i.e., not quantized theory, is not entirely trivial; namely while for $s = 0$, minima of the bosonic part of the action S_5 were simply constants, for $s > 0$ the situation becomes more interesting. In a sense, the Morse function breaks the symmetry that all points of N are equal.

We consider the bosonic part of our action $S_5 + S_{int}$, again on a cylindrical world sheet $\{(t, x) : t \in \mathbb{R}, x \in S^1\}$, and assuming that the fields are independent of x so that we can carry out the x-integration. We then have the total energy or Hamiltonian, see (2.1.7), (2.1.9),

$$H_B(\phi) = \frac{1}{2} L \int dt \left(g_{ij}(\phi) \frac{d\phi^i}{dt} \frac{d\phi^j}{dt} + s^2 g^{ij}(\phi) \frac{\partial h}{\partial \phi^i} \frac{\partial h}{\partial \phi^j} \right). \tag{2.4.131}$$

Obviously, $H_B(\phi) = 0$ if $\phi(t) \equiv q_v$, where q_v is a critical point of h.

These are the classical solutions. We next consider tunneling paths or so-called instanton solutions between such classical solutions.

Given two critical points q_v, q_μ, we have to find

$$\phi : \mathbb{R} \to N; \qquad \lim_{t \to -\infty} \phi(t) = q_v, \qquad \lim_{t \to \infty} \phi(t) = q_\mu \tag{2.4.132}$$

minimizing

$$\begin{aligned}
H_B(\phi) &= \frac{1}{2} L \int dt \left(g_{ij}(\phi) \frac{d\phi^i}{dt} \frac{d\phi^j}{dt} + s^2 g^{ij}(\phi) \frac{\partial h}{\partial \phi^i} \frac{\partial h}{\partial \phi^j} \right) \\
&= \frac{1}{2} L \int dt \left\| \frac{d\phi^i}{dt} \pm s g^{ij} \frac{\partial h}{\partial \phi^j} \right\|^2 \mp s L \int dt \frac{d(h \circ \phi)}{dt} \\
&\geq \pm s L \left(\lim_{t \to -\infty} h(\phi(t)) - \lim_{t \to \infty} h(\phi(t)) \right) \\
&= \pm s L (h(q_v) - h(q_\mu)),
\end{aligned} \tag{2.4.133}$$

(using the simple relation $\|\frac{d\phi^i}{dt} \pm sg^{ij}\frac{\partial h}{\partial \phi^j}\|^2 = g_{ik}(\frac{d\phi^j}{dt} \pm sg^{ij}\frac{\partial h}{\partial \phi^i})(\frac{d\phi^k}{dt} \pm sg^{kl}\frac{\partial h}{\partial \phi^l}))$.
Equality occurs precisely if

$$\frac{d\phi^i}{dt} = \mp sg^{ij}\frac{\partial h}{\partial \phi^j}, \tag{2.4.134}$$

i.e.,

$$\frac{d\phi}{dt} = \mp s(\nabla h)\circ\phi$$

(2.4.134) means that, up to sign, $\phi(t)$ is a curve of steepest descent for h.

Thus, the minimum action paths between any two critical points are paths of steepest descent, and the action of such a path is

$$sL|h(q_v) - h(q_\mu)|.$$

We now let q be a critical point of h of Morse index p, and we let $r_1, \ldots, r_m$ be the critical points of Morse index $p + 1$. We put

$$\delta|q\rangle = \sum_{\mu=1}^{m} n(q, r_\mu)|r_\mu\rangle.$$

Here, we associate to each critical point q of index p a basis vector $|q\rangle$ of a vector space V_p. We put

$$n(q, r_\mu) = \sum_{\Gamma(r_\mu, q)} n_\Gamma$$

where $\Gamma(r_\mu, q)$ is the path of steepest descent from r_μ to q, and where n_Γ is ± 1 according to the following rule.

By the above considerations, each critical point q of index p corresponds to a p-form localized near that point, and this p-form yields an orientation of the subspace of $T_q N$ spanned by the p negative eigendirections of the Hessian of h at q. At r_μ, we thus have a $(p+1)$-form, and in fact the direction of steepest descent corresponds to the eigendirection for the smallest eigenvalue of $\nabla^2 h(r_\mu)$. If we thus transport this $(p + 1)$-form parallely along Γ and contract it with the tangent direction of Γ, we obtain a p-form at q. Comparing the resulting orientation at q with the one coming from the p-form corresponding to q then determines whether n_Γ is $+1$ or -1, i.e., $n_\Gamma = 1$ if they agree, $n_\Gamma = -1$ else.

The important point is that

$$\delta^2 = 0.$$

This can be verified directly or deduced from representing δ as the limit of d_s for $s \to \infty$. (Note that in any case $d_s : V_p \to V_{p+1}$ also yields a coboundary operator, i.e., $d_s^2 = 0$.) It is a standard result of algebraic topology that once one has such a coboundary operator, one obtains the strong Morse inequalities encoded in the formula

$$\sum_p m_p t^p - \sum_p b_p t^p = (1+t)Q(t),$$

where $Q(t)$ is a polynomial with nonnegative integer coefficients.

A more general version of the supersymmetry algebra arises if, in addition to the Hamiltonian H, we also have a momentum operator P, and if the supersymmetry operators Q_1, Q_2 satisfy

$$Q_1^2 = H + P, \qquad Q_2^2 = H - P, \qquad Q_1 Q_2 + Q_2 Q_1 = 0.$$

These relations imply

$$[Q_i, H] = 0 = [Q_i, P] \quad \text{for } i = 1, 2.$$

Also,

$$H = \frac{1}{2}(Q_1^2 + Q_2^2)$$

is again positive semidefinite.

A realization of this supersymmetry algebra arises as follows.

Let X be a Killing field on our compact Riemannian manifold N, i.e., an infinitesimal isometry of N. Let L_X be the Lie derivative in the direction of X, and $i(X)$ the interior multiplication with X of a differential form. For $s \in \mathbb{R}$, we consider

$$d_s = d + si(X).$$

Let d_s^* be the adjoint of d_s. Since X is a Killing field, one computes that

$$d_s^{*2} = -d_s^2.$$

Also

$$d_s^{*2} = -sL_X.$$

The Hamiltonian is

$$H_s = d_s d_s^* + d_s^* d_s.$$

The supersymmetry operators are

$$Q_{1,s} = i^{\frac{1}{2}} d_s + i^{-\frac{1}{2}} d_s^*, \qquad Q_{2,s} = i^{-\frac{1}{2}} d_s + i^{\frac{1}{2}} d_s^*.$$

Defining

$$P = 2isL_X,$$

we then have the above supersymmetry algebra,

$$Q_1^2 = H + P, \qquad Q_2^2 = H - P, \qquad Q_1 Q_2 + Q_2 Q_1 = 0.$$

More generally, one may use a function h invariant under the action of X, i.e.,

$$i(X)dh = 0,$$

and put

$$d_{s_1,s_2} = e^{-hs_2}d_{s_1}e^{hs_2}.$$

Thus, the parameter s_1 corresponds to the Killing field X, whereas s_2 corresponds to the Morse function h. The supersymmetry generators are then

$$Q_{1,s_1,s_2} = i^{\frac{1}{2}}d_{s_1,s_2} + i^{-\frac{1}{2}}d^*_{s_1,s_2},$$

$$Q_{2,s_1,s_2} = i^{-\frac{1}{2}}d_{s_1,s_2} + i^{\frac{1}{2}}d^*_{s_1,s_2}$$

and

$$H_{s_1,s_2} = d_{s_1,s_2}d^*_{s_1,s_2} + d^*_{s_1,s_2}d_{s_1,s_2},$$

$$P = 2is_1L_X.$$

We return to our supersymmetric nonlinear sigma model with a cylindrical world sheet $\mathbb{R} \times S$, where the space S is a circle of circumference L. Instead of considering maps

$$\phi : \mathbb{R} \times S \to N,$$

we may equivalently consider maps

$$\psi : \mathbb{R} \to \Omega_s(N),$$

where $\Omega_s(N)$ is the loop space of maps from S to N. The loop space $\Omega_s(N)$ will now play the role of our target manifold. Of course, in contrast to what was assumed for our target manifold N, $\Omega_s(N)$ is not compact.

The group $U(1)$ of rotations of S acts on $\Omega_s(N)$ by isometries, simply by mapping a loop $\gamma(t)$ to the loop $\gamma(t + a)$ (the addition in S is the one in $\mathbb{R}$ mod L). As before, we may define the operators

$$d_s = d + si(X), \qquad H_s = d_sd^*_s + d^*_sd_s,$$

where X is the generator of the $U(1)$ action.

Of course, due to the fact that $\Omega_s(N)$ is infinite-dimensional, certain problems of convergence arise when trying to carry over the preceding finite-dimensional analysis.

The approach to Morse theory via the supersymmetric sigma model is due to Witten [104, 105]. This in turn led to Floer's approach to Morse theory that constructs the Morse complex from counting flow lines between critical points, see [37] and the expositions in [65, 96].

The supersymmetric action functional (2.4.112) can also be utilized for a proof of the Atiyah–Singer index theorem [7], as discovered by Alvarez-Gaumé [3], Friedan and Windey [41, 42] and Getzler [47, 48]. Systematic expositions can be found in [10] and [43].

2.4.7 The Gravitino

The preceding considerations were local insofar as the supersymmetry variation parameter ε was assumed to be constant. It turns out that from a more global perspective, ε has to be considered to be a section of some bundle and cannot in general be taken to be constant. This implies that also derivatives of ε will enter the supersymmetry computations. We address this issue now and see that it will lead us to very interesting geometric structures.

We start with the linear supersymmetric sigma model from Sect. 2.4.3, that is, the extension of (2.4.3) with a supersymmetric partner for the scalar field ϕ, an anticommuting spinor field ψ:

$$S(\phi, \psi, \Sigma) := \frac{1}{2} \int_\Sigma (\partial_\alpha \phi^a \partial^\alpha \phi_a + \bar{\psi}^a \gamma^\alpha \partial_\alpha \psi_a) \mathrm{d}^2 z. \tag{2.4.135}$$

Here, the $\gamma^\alpha, \alpha = 1, 2$ are standard Dirac matrices, defined by a representation of $Cl(2, 0)$ as above. (Note: In the physics literature, one usually works with a Minkowski world sheet, that is, one takes an indefinite metric on the underlying surface, and consequently considers $Cl(1, 1)$ instead.)

The equations of motion, that is, the Euler–Lagrange equations for (2.4.135) are simple linear equations ((2.4.58), (2.4.59)):

$$\partial^\alpha \partial_\alpha \phi^a = 0, \tag{2.4.136}$$

$$\gamma^\alpha \partial_\alpha \psi^a = 0 \quad \text{for } a = 1, \ldots, d, \tag{2.4.137}$$

that is, ϕ is harmonic and ψ solves the Dirac equation.

Similarly, one can consider a metric g instead of only a conformal structure and consider the functional

$$S(\phi, \psi, g) := \frac{1}{2} \int_S (g^{\alpha\beta} \partial_\alpha \phi^a \partial_\beta \phi_a + \bar{\psi}^a \gamma^\alpha \partial_\alpha \psi_a) \sqrt{\det g} \, \mathrm{d} z^1 \mathrm{d} z^2. \tag{2.4.138}$$

One then has the supersymmetry transformations (2.4.65):

$$\delta \phi^a = \bar{\varepsilon} \psi^a, \tag{2.4.139}$$

$$\delta \psi^a = \gamma^\alpha \partial_\alpha \phi^a \varepsilon \tag{2.4.140}$$

with an anticommuting ε. (Of course, mathematically, one should consider this as a transformation of the independent variables of an underlying superspace instead of as a transformation of the fields.) The commutator of two such transformations yields a spatial translation:

$$[\delta_1, \delta_2] = \delta_1(\bar{\varepsilon}_2 \psi^a) - \delta_2(\bar{\varepsilon}_1 \psi^a) = 2 \bar{\varepsilon}_1 \gamma^\alpha \varepsilon_2 \partial_\alpha \phi^a. \tag{2.4.141}$$

In fact, these are infinitesimal transformations that integrate to local ones, but we also need to consider the global situation. Globally, instead of a translation, we have a diffeomorphism, and so the supersymmetry transformations should generate the superdiffeomorphism group of the underlying supersurface. Also, globally, ε is not a scalar parameter, but transforms as a spin-1/2 field, that is, mathematically, a not necessarily holomorphic, anticommuting section of $K^{1/2}$, K being the canonical bundle of Σ (for some choice of a square root of K, that is, of a spin structure). (Even though, w.r.t. its z-dependence, ε transforms as a section of $K^{1/2}$, it also contains an independent odd parameter; therefore, $\varepsilon\psi = -\psi\varepsilon$, but in general, we do not have $\varepsilon\psi = 0$.)

A supersymmetry transformation induces a variation of S; this is computed as (cf. (2.4.56))

$$\delta S = -2 \int \partial_\alpha \bar{\varepsilon} J^\alpha \qquad (2.4.142)$$

with the **supercurrent**

$$J_\alpha = \frac{1}{2}\gamma^\beta \gamma_\alpha \psi^a \partial_\beta \phi_a. \qquad (2.4.143)$$

Likewise, for a spatial translation, we get the **energy–momentum tensor**:

$$T_{\alpha\beta} = \partial_\alpha \phi^a \partial_\beta \phi_a + \frac{1}{4}\bar{\psi}^a \gamma_\alpha \partial_\beta \psi_a + \frac{1}{4}\bar{\psi}^a \gamma_\beta \partial_\alpha \psi_a - \text{trace}. \qquad (2.4.144)$$

Of course, this is the appropriate generalization of (2.4.9). As before, it is traceless, and again, this can be seen as expressing a (super)conformal invariance. Also, as before, both the supercurrent J and the energy–momentum tensor T are divergence-free when the equations of motion hold. With the same implicit identifications as in Sect. 2.4, T is a holomorphic quadratic differential on Σ, that is, a holomorphic section of K^2, while J is a holomorphic section of $K^{3/2}$.

The preceding facts have several important consequences:

- In line with the general concept of supergeometry, the space of independent variables for the ϕ and ψ fields should be a superspace, that is, here it should be a super Riemann surface (SRS). Then, in the same manner that the Dirichlet integral, the action functional $D(\phi, \Sigma)$, yielded a (co)tangent vector to the moduli space M_p when varying Σ, now variations of Σ for $S(\phi, \psi, \Sigma)$ should yield a (co)tangent vector to the moduli space of super Riemann surfaces. From this, we infer that the tangent space to that space should be given by even holomorphic sections of K^2 and odd holomorphic sections of $K^{3/2}$. In particular, the even dimension should be $3p - 3$ as before while the odd one is $2p - 2$, again by Riemann–Roch.

- As before, our action functional is only invariant on-shell, that is, when J is holomorphic. From (2.4.142), we see the obstruction to global invariance, namely the nonvanishing of $\partial_\alpha \bar{\varepsilon}$. As a spin-1/2 field, ε is a section of a nontrivial bundle and therefore cannot be taken to be globally constant. Thus, the obstruction to full superdiffeomorphism invariance comes from the global topology of the underlying surface.

In order to understand these issues better, we now make the fundamental observation that the functional S from (2.4.135) or (2.4.138) does not yet constitute a full supersymmetric generalization of the functional $S(\phi, g)$ studied in Sect. 2.4. Namely, we have only given ϕ a supersymmetric partner, but not our other field, namely the metric g. We shall do that now and see that this yields a fully satisfactory theory that gives a profound understanding of the moduli space of super Riemann surfaces.

In place of the metric $(g_{\alpha\beta})$, it is convenient to consider a zweibein e_α^a, from which we can reconstruct the metric as $g_{\alpha\beta} = \delta_{ab} e_\alpha^a e_\beta^b$. In other words, we introduce an additional $U(1)$ symmetry which, however, can be easily divided out since that group is compact. The supersymmetric partner of the zweibein is then a gravitino (Rarita–Schwinger field) $\chi_{A\alpha}$ where $A = 1, 2$ is a spinor index that will be suppressed in the sequel, whereas α is a vector index as before. Thus, χ transforms as a spin-3/2 field. This might already suggest how to obtain the moduli space of super Riemann surfaces in analogy to 4 of Sect. 1.4.2. Namely, one would take the space of all metrics (equivalently, after dividing out the $U(1)$ symmetry, zweibeins) and gravitinos, and then divide out all the invariances, that is, the superdiffeomorphisms and superconformal scalings. However, although this idea is conceptually insightful, the actual construction of the moduli space of super Riemann surfaces proceeds differently, see [93].[19] In fact, because the spaces involved, like the one of superdiffeomorphisms, are necessarily infinite-dimensional, Sachse had to replace the standard approach of ringed topological spaces by a categorical reformulation of supergeometry, see [94].

The supersymmetry transformations of the fields ϕ, ψ, e, χ are then

$$\delta\chi_\alpha = \partial_\alpha\bar{\varepsilon}, \tag{2.4.145}$$

$$\delta e_\alpha^a = -2\bar{\varepsilon}\gamma^a\chi_\alpha, \tag{2.4.146}$$

$$\delta\phi^a = \bar{\varepsilon}\psi^a, \tag{2.4.147}$$

$$\delta\psi^a = \gamma^\alpha\varepsilon(\partial_\alpha\phi^a - \bar{\psi}^a\chi_\alpha). \tag{2.4.148}$$

The supersymmetric functional is then

$$
\begin{aligned}
S(\phi, \psi, g, \chi) \\
:= \frac{1}{2}\int_S \Big(& g^{\alpha\beta}\partial_\alpha\phi^a\partial_\beta\phi_a + \bar{\psi}^a\gamma^\alpha\partial_\alpha\psi_a + 2\bar{\chi}_\alpha\gamma^\beta\gamma^\alpha\psi^a\partial_\beta\phi_a \\
& + \frac{1}{2}\bar{\psi}_a\psi^a\bar{\chi}_\alpha\gamma^\beta\gamma^\alpha\chi_\beta\Big)\sqrt{\det g}\,dz^1dz^2.
\end{aligned}
\tag{2.4.149}
$$

[19]Using the zweibeins directly would mean taking the phase space of a 2D supergravity theory as the gauge theory for supersymmetry. One would then in addition need a super connection on Σ whose coefficients are the gauge fields. Dividing out the invariances involved becomes very complicated, and therefore, it is better to proceed as in [93].

Here, the first two terms are those from (2.4.138), the third one is introduced to compensate (2.4.142), and the last one is then needed to compensate the terms coming from the variation of $\partial_\beta \phi$ in the third one.

We have thus obtained a functional that is fully supersymmetric even off-shell.

Summary: We see the merging of a profound mathematical concept, namely that of a moduli space of Riemann surfaces and a deep method from theoretical physics, namely the symmetries of action functionals. This suggests a unique concept of a super Riemann surface, for which we have already described the super moduli space. It remains to be seen how the approaches of Sect. 1.4.2 extend to this setting. Ideally, they should as beautifully coincide as in the situation of ordinary Riemann surfaces.

Of course, the preceding formalism can be recast into the mathematical framework of supergeometry.

We have considered only one of the two supersymmetries arising in string theory, namely world-sheet supersymmetry, but not space–time supersymmetry. The latter refers to the target space, which we have taken to be Euclidean space here. For example, while ψ transforms as a spinor on the domain, it transforms as a vector in the target space. For a discussion of space–time supersymmetry, see, e.g., [50]. Here, instead, we replace the Euclidean target space by a Riemannian manifold N. Equation (2.4.138) then becomes the supersymmetric nonlinear sigma model of quantum field theory as treated in the preceding section. The equations for ϕ and ψ then become nonlinear and coupled, and in fact, ψ is a spinor-valued section of $\phi^* TN$, the pull-back of the tangent bundle of N under the map ϕ. Naturally, one can also include the fields g and χ into these considerations, by expanding not only with respect to the map into N, but also with respect to the domain metric.

The supersymmetric action functional with gravitino term is discussed in [26, 50], with more details in [27]. The moduli space of super Riemann surfaces has been constructed from the global analysis perspective advocated here by Sachse [93].

2.5 Functional Integrals

We can now bring the material of the preceding sections together and discuss general (Gaussian) functional integrals. These are formal integrals of the form

$$\int D\varphi \, e^{-S(\varphi)} \tag{2.5.1}$$

where $S(\varphi)$ is some quadratic Lagrangian action as introduced in Sect. 2.2 and we formally integrate w.r.t. to some collection of fields φ. We can, of course, also introduce Planck's constant and replace (2.5.1) by

$$\int D\varphi \, e^{-\frac{1}{\hbar}S(\varphi)}. \tag{2.5.2}$$

When we consider the heuristic limit $\hbar \to 0$, we see that the minimizers of the action S dominate the functional integral more and more, because other fields φ yield exponentially smaller contributions. For physicists, it is then natural to perform an expansion of (2.5.2) in terms of $\frac{1}{\hbar}$, the so-called stationary phase approximation.

Certainly, one can also consider the oscillatory integral

$$\int D\varphi \, e^{\frac{i}{\hbar}S(\varphi)}, \tag{2.5.3}$$

which we may view as a generalization of the Feynman path integral discussed in Sect. 2.1.3.

As before, see Sects. 2.1.2, 2.1.3, we consider Gaussian functional integrals as formal analogs of Gaussian integrals with infinitely many variables. In addition, we shall make use of the invariance considerations in Sect. 2.3.2 to divide out symmetries.

There is one general issue that can be contemplated at this point: It is a general principle of quantum field theory that no arbitrary choices are permitted. Whenever something is selected from some class of possibilities, one should integrate out the possible values of the selection, weighted with some (negative or imaginary) exponential of the underlying action. Thus, we consider (2.5.1) when we have a collection of fields φ. After normalization, we consider $\frac{1}{Z}D\varphi \, e^{-S(\varphi)}$ (where the constant Z has been chosen so that the total integral of the measure becomes 1) as a probability measure on the space of fields (similar to a Gibbs measure in statistical mechanics). For any function $f(\varphi)$ of the field φ, we can then compute its expectation value as

$$\frac{1}{Z}\int D\varphi \, f(\varphi) e^{-\frac{1}{\hbar}S(\varphi)}. \tag{2.5.4}$$

In mathematics, instead of taking a functional integral, in the situation where some underlying structure has to be selected, one attempts to equip the space of all possible choices with some geometric structure. That is then called a moduli space. Above, we have discussed the moduli space of Riemann surfaces.

2.5.1 Normal Ordering and Operator Product Expansions

The following example will bring out the essential aspects. Let (M, g) be a compact Riemannian manifold of dimension d. For a function φ on M, we put

$$S(\varphi) = \frac{1}{4\pi\alpha'}\int_M (\|D\varphi\|^2 + m^2\varphi^2) \, d\mathrm{Vol}_g(M)$$

$$= \frac{1}{4\pi\alpha'}(\varphi, (-\Delta_g + m^2)\varphi)_{L^2}, \tag{2.5.5}$$

where α' is a constant, the so-called Regge slope. Thus, this is essentially the same functional as the one considered in Sect. 2.4, see (2.4.3), with the difference that here we have an additional mass term and a different normalization factor in front of the integral. $\Delta = \Delta_g$ is the Laplace–Beltrami operator of (M, g), defined in (1.1.103), (2.4.4).

We note some differences here compared to Sect. 2.1.3. There, we had taken functional integrals for paths x (in some Euclidean or Minkowski space), that is, mappings $x : [t', t''] \to \mathbb{R}^d$, say, with fixed boundary conditions $x(t') = x'$, $x(t'') = x''$. Here, we are integrating functions over a more general domain, namely a Riemannian manifold, and we do not impose boundary conditions. In fact, M may be some closed manifold without boundary. If M does have a boundary, we can also impose a boundary condition via an insertion into our functional integral.

According to the general scheme just discussed, the choice of the manifold (M, g) represents an arbitrary choice, and therefore, one should integrate out all such choices, that is, take another functional integral w.r.t. all possible metrics g on M, and perhaps also a sum w.r.t. all diffeomorphism types of M. That is, in fact, done in string theory, where the dimension of M is fixed to be 2 and one then formally integrates w.r.t. all metrics and sums with respect to the different genera of the underlying surface.

We also consider the propagator of the free field of mass m, or, in mathematical terminology, the Green operator

$$G = 2\pi \alpha'(-\Delta + m^2)^{-1}. \tag{2.5.6}$$

Thus,

$$S(\varphi) = \frac{1}{2}(\varphi, G^{-1}\varphi)_{L^2}. \tag{2.5.7}$$

The fundamental object of interest is the partition function (in older texts, this is denoted by the German term *Zustandssumme*)

$$Z := \int D\varphi \exp(-S(\varphi))$$

$$= \int D\varphi \exp\left(-\frac{1}{2}(\varphi, G^{-1}\varphi)\right) \tag{2.5.8}$$

with a formal integration over all functions $\varphi \in L^2(M)$.

The analogy with the above discussion of Gaussian integrals (2.1.24), obtained by replacing the coordinate index i in (2.1.24) by the point z in our manifold M, would suggest

$$Z = (\det G)^{\frac{1}{2}}, \tag{2.5.9}$$

when we normalize

$$D\varphi = \prod_i \frac{d\varphi_i}{\sqrt{2\pi}} \tag{2.5.10}$$

to get rid of the factor $(2\pi)^n$ in (2.1.24). Here, the (φ_i) are an orthonormal basis of the Hilbert space $L^2(M)$, for example, the eigenfunctions of Δ.

The idea is then to define $\det G$ as the renormalized product of the eigenvalues λ_n of G. The mathematical construction is based on the Weyl estimates. By these estimates, the λ_n behave as $O(n^{-\frac{2}{d}})$. Motivated by (2.1.26), one then puts

$$\det G := \exp(-\zeta_G'(0)), \qquad (2.5.11)$$

where the ζ-function $\zeta_G(s)$ is the meromorphic continuation of $\sum_n \lambda_n^{-s}$, defined for $\mathrm{Re}(s) < -\frac{d}{2}$, to the entire complex plane; it is analytic at 0. This procedure is called zeta function regularization. The determinant defined by (2.5.11) has certain multiplicative properties like the ordinary determinant, see e.g. [111].

Comparing (2.5.8) with (2.1.24), the analogy is then that the coordinate values $x^1, \ldots, x^d$ get replaced by the values of the function φ at the points $y \in M$. That is, we have infinitely many degrees of freedom, corresponding to the points $y \in M$ instead of to the discrete indices $i = 1, \ldots, n$. The values of these degrees of freedom are then assembled into the function φ in place of the vector $x = (x^1, \ldots, x^n)$.

In analogy with (2.1.30), for points $y_1, \ldots, y_m \in M$, we may then define correlation functions

$$\langle \varphi(y_1) \cdots \varphi(y_m) \rangle := \frac{1}{Z} \int D\varphi \varphi(y_1) \cdots \varphi(y_m) \exp(-S(\varphi)). \qquad (2.5.12)$$

Note that, in contrast to Sect. 2.1.3, here we are normalizing the integrals by dividing by Z, so that these correlation functions can be interpreted as the expectation values of the product of the evaluations of the fields at the points $y_1, \ldots, y_m$. Again, these vanish for odd m (because a Gaussian integral is quadratic in the fields, hence even), and as in (2.1.31)

$$\langle \varphi(y_1) \varphi(y_2) \rangle = G(y_1, y_2). \qquad (2.5.13)$$

Here, the Green function $G(y_1, y_2)$ is the kernel of the operator G, and it has a singularity at $y_1 = y_2$, of order $\log \mathrm{dist}(y_1, y_2)$ for $d = 2$ and $\mathrm{dist}(y_1, y_2)^{2-d}$ for $d > 2$.

Likewise, the analog of Wick's theorem (2.1.32) holds.

We now specialize to the case where the particle is massless, i.e., $m = 0$ in (2.5.5), and M is a Riemann surface Σ. Thus, in complex coordinates, the action is

$$S = \frac{1}{2\pi\alpha'} \int d^2 w \, \partial\varphi \bar{\partial}\varphi. \qquad (2.5.14)$$

We note that the metric g here disappears from the picture. This comes from the fact that S in (2.5.14) is conformally invariant, that is, remains unchanged when the underlying metric is multiplied by some positive function, and therefore depends only on the conformal structure, that is, on the Riemann surface on which it is defined, but not on a particular choice of a conformal metric on that Riemann surface. The issue of conformal invariance plays a fundamental role in conformal field theory and string theory, see [26, 46, 62].

The classical equation of motion is (2.4.21),

$$\partial\bar{\partial}\varphi(z, \bar{z}) = 0. \qquad (2.5.15)$$

One writes the argument here as $(z, \bar{z})$ instead of simply z, because the notation $f(z)$ is reserved for a holomorphic function, as explained in Sect. 1.1.2. (2.5.15) implies that $\partial \varphi$ is a holomorphic function $\partial \varphi(z)$, and $\bar{\partial} \varphi$ is an antiholomorphic function $\bar{\partial} \varphi(\bar{z})$.

The complex coordinates

$$z = x^1 + ix^2,$$

$$\bar{z} = x^1 - ix^2$$

admit a Minkowski continuation with $x^0 = -ix^2$. Then, a holomorphic function is a function of $x^0 - x^1$, an antiholomorphic one is a function of $x^0 + x^1$. One calls an (anti)holomorphic function left-(right-)moving.

As before, we wish to compute the expectation values

$$\langle F(\varphi) \rangle = \frac{1}{Z} \int D\varphi \exp(-S) F(\varphi). \tag{2.5.16}$$

We shall now repeat some of the discussion of Sect. 2.1.3 and see how it applies to the present situation. The above analogy between ordinary integrals and path integrals said that the finitely many ordinary degrees of freedom, the coordinate values of the integration variable, are replaced by the infinitely many function values $\varphi(z, \bar{z})$. Therefore, integration by parts should yield that

$$0 = \int D\varphi \frac{\delta}{\delta \varphi(z, \bar{z})} \exp(-S). \tag{2.5.17}$$

This gives

$$0 = -\int D\varphi \exp(-S) \frac{\delta S}{\delta \varphi(z, \bar{z})},$$

and so,

$$0 = -\left\langle \frac{\delta S}{\delta \varphi(z, \bar{z})} \right\rangle = \frac{1}{\pi \alpha'} \langle \partial \bar{\partial} \varphi(z, \bar{z}) \rangle. \tag{2.5.18}$$

Thus, the classical equation of motion (2.5.15) becomes an equation for the expectation value of the corresponding operator. Equation (2.5.18) can also be written as

$$\frac{1}{\pi \alpha'} \partial_z \partial_{\bar{z}} \langle \varphi(z, \bar{z}) \rangle = 0. \tag{2.5.19}$$

Let us return to (2.5.16). The functional $F(\varphi)$ typically represents certain linear combinations of products of evaluations of φ at points $z_1, \ldots, z_m \in \Sigma$. When none of those points coincides with the point z for which we take the functional derivative $\frac{\delta}{\delta \varphi(z, \bar{z})}$, the preceding computation also goes through for $F(\varphi)$.

Things change when one of those insertion points is allowed to coincide with z. In Sect. 2.1.3, that led us to the temporal ordering scheme for operators. Similarly, here, from the analog of (2.1.125), we shall be led to the so-called normal ordering scheme.

For example

$$0 = \int D\varphi \frac{\delta}{\delta\varphi(z,\bar{z})} (\exp(-S)\varphi(\zeta,\bar{\zeta}))$$

$$= \int D\varphi \exp(-S)\left(\delta(z-\zeta,\bar{z}-\bar{\zeta}) + \frac{1}{\pi\alpha'}(\partial_z\partial_{\bar{z}}\varphi(z,\bar{z}))\varphi(\zeta,\bar{\zeta})\right). \qquad (2.5.20)$$

Thus

$$0 = \left\langle \delta(z-\zeta,\bar{z}-\bar{\zeta}) + \frac{1}{\pi\alpha'}\partial_z\partial_{\bar{z}}\varphi(z,\bar{z})\varphi(\zeta,\bar{\zeta})\right\rangle. \qquad (2.5.21)$$

Again, this is not affected by other insertions not coincident with z.

We thus interpret (2.5.18) and (2.5.21) as operator equations, that is, as holding for all components of the corresponding quantum mechanical operators, since these are precisely obtained by such insertions.

We thus write the operator equation

$$\frac{1}{\pi\alpha'}\partial_z\partial_{\bar{z}}\varphi(z,\bar{z})\varphi(\zeta,\bar{\zeta}) = -\delta(z-\zeta,\bar{z}-\bar{\zeta}), \qquad (2.5.22)$$

as in (2.1.125). When we solve (2.5.22), we therefore obtain a Green function type singularity, $\log|z-\zeta|^2$.

In order to eliminate this contribution, one considers the normal ordered operators

$$:\varphi(z,\bar{z}): = \varphi(z,\bar{z}),$$

$$:\varphi(z_1,\bar{z}_1)\varphi(z_2,\bar{z}_2): = \varphi(z_1,\bar{z}_1)\varphi(z_2,\bar{z}_2) + \frac{\alpha'}{2}\log|z_1-z_2|^2. \qquad (2.5.23)$$

This quantum correction will below lead to a central extension of the Lie algebra of the diffeomorphism group of the circle (see (2.5.63), (2.5.66) in Sect. 2.5.3).

We then have

$$\partial_1\bar{\partial}_1:\varphi(z_1,\bar{z}_1)\varphi(z_2,\bar{z}_2): = 0. \qquad (2.5.24)$$

Thus, $:\varphi(z_1,\bar{z}_1)\varphi(z_2,\bar{z}_2):$ is a harmonic function and therefore locally the sum of a holomorphic and an antiholomorphic function. From this, we obtain the Taylor expansion

$$\varphi(z_1,\bar{z}_1)\varphi(z_2,\bar{z}_2)$$

$$= -\frac{\alpha'}{2}\log|z_1-z_2|^2$$

$$+ \sum_{\nu=1}^{\infty}\frac{1}{\nu!}\left((z_1-z_2)^{\nu}:\varphi\partial^{\nu}\varphi(z_2,\bar{z}_2): + (\bar{z}_1-\bar{z}_2)^{\nu}:\varphi\bar{\partial}^{\nu}\varphi(z_2,\bar{z}_2):\right) \qquad (2.5.25)$$

(mixed terms with $\partial\bar{\partial}$ vanish by the equation of motion; note that in general, derivatives need not commute with normal ordering).

Equation (2.5.25) is the prototype of an operator product expansion (OPE). As discussed, the φ's here are considered as quantum mechanical operators.

The transition from functions to operators needs some explanation. In (2.5.16), we can add insertions $I^1, \ldots, I^m$, that is, functions of φ evaluated at some points $z_1, \ldots, z_m \in M$. Thus, we have expressions of the form

$$\frac{1}{z} \int D\varphi \exp(-S) F(\varphi) I^1(\varphi)(z_1, \bar{z}_1) \cdots I^m(\varphi)(z_m, \bar{z}_m).$$

More generally, we can also have insertions of the form

$$\int I(z, \bar{z}; \varphi) \, d\mu(z)$$

for some measure $d\mu(z)$. For example, these insertions can be certain boundary conditions represented by Dirac functionals. When we do not specify these insertions, we simply write

$$\langle F(\varphi) \cdots \rangle.$$

$F(\varphi)$ then determines an operator $\hat{F}(\varphi)$ operating on such insertions. $\langle F(\varphi) \rangle$ is the matrix element $\langle 0|\hat{F}(\varphi)|0 \rangle$ of $\hat{F}(\varphi)$ where $|0\rangle$ is the vacuum.

2.5.2 Noether's Theorem and Ward Identities

Before proceeding, we need to translate Noether's theorem into the operator setting. The result is a Ward identity.

As in Sect. 2.3.2, we consider a general Lagrangian action

$$S = \int F(\varphi(x), d\varphi(x)) \, dx \qquad (2.5.26)$$

and transformations

$$x \mapsto x',$$

$$\varphi(x) \mapsto \varphi'(x') =: \psi(\varphi(x)).$$

Infinitesimally,

$$x' = x + s\delta x, \qquad (2.5.27)$$

$$\varphi'(x') = \varphi(x) + s\delta\psi(x). \qquad (2.5.28)$$

By (2.3.28), the Noether current is

$$j_i^\alpha = \left(-F_{p\alpha} \frac{\partial \varphi}{\partial x^\beta} + \delta_\beta^\alpha F\right) \frac{\delta x^\beta}{\delta s_i} + F_{p\alpha} \frac{\delta \psi}{\delta s_i}, \qquad (2.5.29)$$

$$\delta S = -\int dx j_i^\alpha \partial_\alpha s_i$$

$$= \int dx \partial_\alpha j_i^\alpha s_i. \qquad (2.5.30)$$

According to Noether (2.3.29), invariance implies a conserved current:

$$\partial_\alpha j_i^\alpha = 0. \qquad (2.5.31)$$

We now turn to the quantum version, that is, invariance of correlation functions, when action and functional integral measure both are invariant:

$$\langle \varphi(x_1') \cdots \varphi(x_n') \rangle = \langle \psi(\varphi(x_1)) \cdots \psi(\varphi(x_n)) \rangle \qquad (2.5.32)$$

by renaming variables ($\varphi \mapsto \varphi'$) and transforming $D\varphi'$ to $D\varphi$.

Ward identities express symmetries in QFT as identities between correlation functions. According to (2.3.27), the field variations are given by

$$G\varphi := \frac{\delta \psi}{\delta s} - \frac{\delta x}{\delta s} \frac{\partial \varphi}{\partial x}. \qquad (2.5.33)$$

For a collection $\Phi = \varphi(x_1) \cdots \varphi(x_n)$ of fields, we have by invariance

$$\frac{1}{Z} \int D\varphi \Phi \exp(-S(\varphi))$$

$$= \langle \Phi \rangle$$

$$= \frac{1}{Z} \int D\varphi' (\Phi + \delta\Phi) \exp\left(-\left(S(\varphi) + \int dx \partial_\alpha j_i^\alpha s_i(x)\right)\right).$$

If the measure is invariant, i.e., $D\varphi' = D\varphi$, then by differentiating w.r.t. s gives

$$\langle \delta\Phi \rangle = \int dx \partial_\alpha \langle j_i^\alpha(x)\Phi \rangle s_i(x) \qquad (2.5.34)$$

(note that Φ does not depend explicitly on x, and thus $\partial(j(x))\Phi = \partial(j(x)\Phi)$).

Since

$$\delta\Phi = -\sum_{k=1}^n (\varphi(x_1) \cdots G\varphi(x_k) \cdots \varphi(x_n)) s(x_k)$$

$$= \int dx s(x) \sum_{k=1}^n (\varphi(x_1) \cdots G\varphi(x_k) \cdots \varphi(x_n)) \delta(x - x_k),$$

we obtain the Ward identity for the current j:

$$\frac{\partial}{\partial x^\alpha}\langle j^\alpha(x)\varphi(x_1)\cdots\varphi(x_n)\rangle = \sum_{k=1}^{n}\langle\varphi(x_1)\cdots G\varphi(x_k)\cdots\varphi(x_n)\rangle\delta(x-x_k). \quad (2.5.35)$$

We now assume that the time $t = x_1^0$ is different from all the times $x_2^0,\ldots,x_n^0$ occurring in Φ. We integrate (2.5.35) between $t-\varepsilon$ and $t+\varepsilon$ for small $\varepsilon > 0$ to obtain

$$\langle Q(t+\varepsilon)\varphi(x_1)\Phi\rangle - \langle Q(t-\varepsilon)\varphi(x_1)\Phi\rangle = \langle G\varphi(x_1)\Phi\rangle \quad (2.5.36)$$

for the charge Q (defined as in (2.3.40)). When we time order the operators, we need to exchange $Q(t-\varepsilon)$ and $\varphi(x_1)$, because $t-\varepsilon < t = x_1^0$. Since (2.5.36) holds for any such Φ, we obtain

$$[Q, \varphi] = G\varphi. \quad (2.5.37)$$

Thus, the conserved change Q is the infinitesimal generator of the symmetry transformations in the operator formalism.

If instead of a Minkowski space–time, we consider Euclidean space, the time ordering is replaced by a radial ordering of the operators as will be discussed in Sect. 2.5.3 below.

2.5.3 Two-dimensional Field Theory

We now compare the preceding with 2-dimensional field theory. We have a spatial coordinate w^1 that may be bounded or periodic,

$$w^1 \sim w^1 + 2\pi, \quad (2.5.38)$$

and a Euclidean time coordinate $\tau = w^2$,

$$-\infty < w^2 < \infty. \quad (2.5.39)$$

We put

$$w = w^1 + iw^2 \quad \text{(equal time coordinates are horizontal lines)} \quad (2.5.40)$$

and

$$z = e^{-iw} \quad \text{(equal time coodinates are concentric circles } C \text{ about}$$
$$\text{origin } z = 0 \text{ which corresponds to the infinite past } w^2 = -\infty). \quad (2.5.41)$$

When going from the Minkowski coordinates w^1, w^2 to the complex coordinates z, temporal invariance $w^2 \to w^2 + t$ then becomes radial invariance $z \to \lambda z$ with

$\lambda \in \mathbb{R}$. This is the starting point of conformal invariance and constitutes one motivation for conformal field theory below.

In the z-coordinates, the charges Q (2.3.40) then become contour integrals of currents j

$$Q\{C\} = \oint_C \frac{dz}{2\pi i} j. \tag{2.5.42}$$

Here, we assume that the current j is meromorphic, without poles on the contour C, of course.

We now consider

$$Q_1\{C_1\}Q_2\{C_2\} - Q_2\{C_2\}Q_1\{C_3\}. \tag{2.5.43}$$

This corresponds to a time ordering $\tau_1 > \tau_2 > \tau_3$.

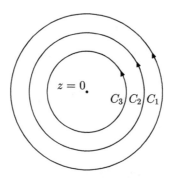

Therefore, when we time order the operators $\hat{Q}_i$ corresponding to the Q_i, we obtain the expression

$$\hat{Q}_1\hat{Q}_2 - \hat{Q}_2\hat{Q}_1 = [\hat{Q}_1, \hat{Q}_2]. \tag{2.5.44}$$

We now consider a point $z_2 \in C_2$, and we can deform the contours as follows:

When we consider infinitesimal time differences, $\tau_1 = \tau_2 + \varepsilon$, $\tau_3 = \tau_2 - \varepsilon$, $\varepsilon \to 0$, we contract the contour $C_1 - C_3$ to C_2, that is, the small circle about z_2, to the point z_2.

We obtain from (2.5.42)–(2.5.44), leaving out the $\hat{}$ for the operators as usual,

$$[Q_1, Q_2]\{C_2\} = \oint_{C_2} \frac{dz_2}{2\pi i} \mathrm{Res}_{z_1 \to z_2} j_1(z_1) j_2(z_2). \tag{2.5.45}$$

This is a fundamental relation. On the l.h.s., we have the commutator algebra of the charges, while on the r.h.s., the singular terms in the operator product expansions (OPEs) of the currents appear.

Instead of the conserved charge $Q_2\{C_2\}$, we can also take an operator $A(z, \bar{z})$ to obtain

$$[Q, A(z_2, \bar{z}_2)] = \text{Res}_{z_1 \to z_2} j(z_1) A(z_2, \bar{z}_2). \tag{2.5.46}$$

When Q is the conserved charge for a variation δ, as in Sect. 2.3.2,

$$\varphi(x) \mapsto \varphi(x) + i\varepsilon s(x), \tag{2.5.47}$$

we have, by (2.5.37),

$$[Q, A(z, \bar{z})] = -\frac{1}{i\varepsilon}\delta A(z, \bar{z}). \tag{2.5.48}$$

(2.5.46) and (2.5.48) yield

$$\text{Res}_{z_1 \to z_2} j(z_1) A(z_2, \bar{z}_2) = -\frac{1}{i\varepsilon}\delta A(z_2, \bar{z}_2). \tag{2.5.49}$$

We now consider a conserved current j in a two-dimensional field theory. As a conserved current, by (2.3.22) and (2.3.29), it is divergence free, that is

$$\partial_{\bar{z}} j_z + \partial_z j_{\bar{z}} = 0. \tag{2.5.50}$$

Taking as a model the energy–momentum tensor T in Sect. 2.4, we now assume that we have

$$j_z = \delta z j_{zz} + \delta\bar{z} j_{z\bar{z}}, \qquad j_{\bar{z}} = \delta z \ j_{\bar{z}z} + \delta\bar{z} \ j_{\bar{z}\bar{z}} \tag{2.5.51}$$

for some holomorphic variation δz. Equation (2.5.50) then becomes

$$\partial_{\bar{z}} j_{zz} + \partial_z j_{\bar{z}z} = 0,$$
$$\partial_z j_{\bar{z}\bar{z}} + \partial_{\bar{z}} j_{z\bar{z}} = 0. \tag{2.5.52}$$

We also assume that the tensor $(j_{zz}, \ldots)$ is symmetric:

$$j_{z\bar{z}} = j_{\bar{z}z}, \tag{2.5.53}$$

and (noting that $\text{tr} \ j = g^{ab} j_{ba} = g^{z\bar{z}} j_{\bar{z}z} + g^{\bar{z}z} j_{z\bar{z}}$) trace-free:

$$j_{z\bar{z}} = 0, \tag{2.5.54}$$

which it has to be for the theory to be conformally invariant.

These relations imply that it is holomorphic:

$$\partial_{\bar{z}} j_{zz} = 0,$$
$$\partial_z j_{\bar{z}\bar{z}} = 0. \tag{2.5.55}$$

We put

$$j(z) = j_{zz}(z),$$

$$\bar{j}(z) = j_{\bar{z}\bar{z}}(\bar{z}).$$

This implies that $f(z)j(z)$ is conserved as well:

$$\partial_{\bar{z}}(fj) = 0, \tag{2.5.56}$$

for any holomorphic function $f(z)$.

Thus, we obtain infinitely many conserved currents. In two-dimensional field theory, this corresponds to the fact that the local conformal group is infinite-dimensional, as conformal invariance led to the energy–momentum tensor T as our conserved current j in Sect. 2.4.

For each holomorphic f, we therefore obtain a conserved charge

$$Q_f = \oint_C \frac{dz}{2\pi i} f(z)T(z) \tag{2.5.57}$$

which generates the conformal transformation

$$z \mapsto z + \varepsilon f(z). \tag{2.5.58}$$

According to (1.1.89), the induced transformation of a field $\varphi(z, \bar{z})$ is

$$\varphi(z, \bar{z}) \mapsto \varphi(z, \bar{z}) + \delta_{f, \bar{f}}\varphi(z, \bar{z}) \tag{2.5.59}$$

with

$$\delta_{f, \bar{f}}\varphi(z, \bar{z}) = (h\partial_z f + \tilde{h}\partial_{\bar{z}}\bar{f} + f\partial_z + \bar{f}\partial_{\bar{z}})\varphi(z, \bar{z}) \tag{2.5.60}$$

where h and $\tilde{h}$ are the conformal weights of φ.

We consider an $(h, 0)$-form $\varphi(z, \bar{z})(dz)^h$. Equations (2.5.48), (2.5.49) and (2.5.57) give

$$\delta_f\varphi(z) = -[Q_f, \varphi(z)]$$

$$= -\mathrm{Res}_{z_1 \to z_2} f(z_1)T(z_1)\varphi(z)$$

$$= \oint_C \frac{dz_1}{2\pi i} f(z_1)T(z_1)\varphi(z), \tag{2.5.61}$$

where C is now a small circle about z.

Since $\tilde{h} = 0$ here, we obtain from (2.5.60) and (2.5.61) that

$$T(z_1)\varphi(z) = \frac{h\varphi(z)}{(z_1 - z)^2} + \frac{\partial_z\varphi(z)}{z_1 - z} + \text{finite terms}. \tag{2.5.62}$$

In particular, for an $(h, 0)$-form, the conformal weight h can be recovered from the operator product with the energy–momentum tensor.

If we take, instead of φ, the energy–momentum tensor T itself in this OPE, we obtain an additional term that essentially comes from the fact that T involves a square of derivatives of fields which induce additional commutator terms:

$$T(z_1)T(z) = \frac{c}{2(z_1 - z)^4} + \frac{2T(z)}{(z_1 - z)^2} + \frac{\partial_z T(z)}{z_1 - z} + \text{finite terms.} \qquad (2.5.63)$$

Here, c is some constant, the so-called central charge. Since T is holomorphic, we can Laurent-expand it:

$$T(z) = \sum_{m=-\infty}^{\infty} \frac{L_m}{z^{m+2}}, \qquad (2.5.64)$$

that is,

$$L_m = \oint_{C_0} \frac{dz}{2\pi i} z^{m+1} T(z) \qquad (2.5.65)$$

for a circle C_0 about the origin $z = 0$.

The L_m are the generators of the Virasoro algebra

$$[L_n, L_m] = \oint_{C_0} \frac{dz}{2\pi i} \oint_{C_z} \frac{dz_1}{2\pi i} z_1^{m+1} z^{m+1} \left[\frac{c}{2(z_1 - z)^4} + \frac{2T(z)}{(z_1 - z)^2} + \frac{\partial T(z)}{z_1 - z} \right]$$

$$= \frac{c}{12} n(n-1)(n+1)\delta_{m+n} + (n-m)L_{m+n}. \qquad (2.5.66)$$

To obtain this, one uses

$$z_1^{n+1} = ((z_1 - z) + z)^{n+1}$$

$$= \frac{n^3 - n}{6}(z_1 - z)^3 z^{n-2} + \frac{n^2 + n}{2}(z_1 - z)^2 z^{n-1}$$

$$+ (n+1)(z_1 - z)z^n + z^{n+1} + \cdots .$$

Summary: The generators of the Virasoro algebra are the Laurent coefficients of the energy–momentum tensor T. The expansion comes from the holomorphicity of T, which in turn follows from the invariance properties of CFT. Since, in contrast to the classical action, the quantum expectation values are not conformally invariant, we obtain a central charge $c \neq 0$ in the commutators of the L_m.

L_0, L_1 and L_{-1} generate an algebra isomorphic to $\mathfrak{sl}(2, \mathbb{R})$, the Lie algebra of $Sl(2, \mathbb{R})$. That Lie algebra is represented here by infinitesimal transformations of the form $\alpha + \beta z + \gamma z^2 = \delta z$, the infinitesimal version at $a = d = 1, b = c = 0$ of $z \mapsto \frac{a z + b}{c z + d}$, the operation of $Sl(2, \mathbb{R})$. In fact, for $n, m = -1, 1, 0$, (2.5.66) is the same as (1.3.48), except for the different notation, of course. In general, L_n generates $\delta z = z^{n+1}$. L_n acts on a primary field (*primary* can be defined by this relation) as

$$[L_n, \varphi(z)] = z^n (z\partial_z + (n+1)h) \varphi(z). \qquad (2.5.67)$$

There is one point here that needs clarification, the relationship between the classical energy–momentum tensor as defined in Sect. 2.4, see (2.4.8), and its operator version. According to (2.4.8), the energy–momentum tensor is the Noether current associated with a variation of the inverse metric $\gamma^{\alpha\beta}$:

$$\delta S = \int dx T_{\alpha\beta}\delta\gamma^{\alpha\beta}. \tag{2.5.68}$$

Quantum mechanically, we have

$$Z_{\gamma+\delta\gamma} = \int (D\varphi)_{\gamma+\delta\gamma}\exp(-S(\varphi,\gamma+\delta\gamma))$$

$$= \int (D\varphi)_\gamma\left(1 + \int dx T\delta\gamma^{-1}\right)\exp(-S(\varphi,\gamma)),$$

assuming that the energy–momentum tensor incorporates both the variation of the action and of the measure,

$$= Z_\gamma + Z_\gamma\int dx\delta\gamma^{-1}\langle T\rangle_\gamma.$$

Thus,

$$\frac{1}{Z_\gamma}\delta Z_\gamma = \int dx\delta\gamma^{-1}\langle T\rangle_\gamma, \tag{2.5.69}$$

or, putting in a factor of 4π for purposes of normalization,

$$\frac{1}{Z_\gamma}\frac{\delta}{\delta\gamma^{-1}(y)}Z_\gamma = \frac{1}{4\pi}\langle T(y)\rangle_\gamma, \tag{2.5.70}$$

and more generally,

$$\frac{1}{Z_\gamma}\frac{(4\pi)^m\delta^m}{\delta\gamma^{-1}(y_1)\cdots\delta\gamma^{-1}(y_m)}(Z_\gamma\langle\varphi(x_1)\cdots\varphi(x_n)\rangle)$$

$$= \langle T(y_1)\cdots T(y_m)\varphi(x_1)\cdots\varphi(x_n)\rangle. \tag{2.5.71}$$

The variation of the measure then induces the central charge c in the expansion (2.5.63) of the operator version of the energy–momentum tensor.

2.6 Conformal Field Theory

2.6.1 Axioms and the Energy–Momentum Tensor

Conformal field theory was introduced by several people. An early paper that was important for the subsequent development of the theory is [9]. A monograph devoted to this topic is [38]. We shall also utilize the treatments in [77] and [46].

In the preceding, we have derived certain formal consequences of the functional integral (2.5.5). In particular, the partition function and the correlation functions satisfy certain relations, and from those, we have obtained the energy–momentum tensor. Its classical version could be identified with a holomorphic quadratic differential in Sect. 2.4. One problem, however, was the definition of the functional integral (2.5.5). There, we briefly discussed the mathematical definition in terms of zeta functions, see (2.5.11), and the spectrum of the Laplace–Beltrami operator. One way to circumvent that problem is to take the indicated algebraic relations and holomorphicity properties as the starting point for an axiomatic theory. This is the idea of conformal field theory.

Thus, abstract conformal field theory specifies for each Riemann surface Σ with a metric g a partition function Z_g and correlation functions $\langle \varphi_1(x_1) \cdots \varphi_n(x_n) \rangle$ for the primary fields with non-coincident $x_1, \ldots, x_n$. These basic data do not need any action or functional integral—although (2.5.5) remains a prime example. The theory is defined in terms of symmetry properties of these correlation functions.

Essentially, these are:

(i) Diffeomorphism covariance: for a diffeomorphism $k : \Sigma \to \Sigma$,

$$Z_g = Z_{k^*g}, \tag{2.6.1}$$

$$\langle \varphi_1(k(x_1)) \cdots \varphi_n(k(x_n)) \rangle_g = \langle \varphi_1(x_1) \cdots \varphi_n(x_n) \rangle_{k^*g}. \tag{2.6.2}$$

(ii) Local conformal covariance

$$Z_{e^\sigma g} = \exp\left(\frac{c}{96\pi} \left(\|d\sigma\|_{L^2_g}^2 + 4 \int_\Sigma \sigma(x) R(x) \right) \right) Z_g, \tag{2.6.3}$$

$$\langle \varphi_1(x_1) \cdots \varphi_n(x_n) \rangle_{e^\sigma g} = \prod_{i=1}^{n} \exp(-h_i \sigma(x_i)) \langle \varphi_1(x_1) \cdots \varphi_n(x_n) \rangle_g. \tag{2.6.4}$$

Here, $R(x)$ is the scalar curvature of (Σ, g), and h_i is the conformal weight (see below) of the field φ_i, as introduced in Sect. 1.1.2; c is called the central charge of the theory. (For the conformal field theory defined by (2.5.5), we have $c = 1$.)

In particular, and this is the fundamental point, the quantum mechanical partition function is not conformally invariant, but instead transforms with a certain factor that depends on the central charge.

We return to the formula (2.5.9) for the functional (2.5.14) on a Riemann surface for $m = 0$. Since $G = 2\pi\alpha'(-\Delta)^{-1}$, we should have, up to a factor,

$$\det G = (\det \Delta)^{-1}.$$

Since, however, Δ has the eigenvalue 0 ($\Delta\varphi_0 = 0$ for a constant function φ_0), we need to restrict it to the orthogonal complement of the kernel of Δ, that is, to the L^2-functions φ with $\int_M \varphi dvol_g(M) = 0$, when defining the determinant by ζ-function regularization. The corresponding determinant is denoted by $\det'$. In fact, one should also normalize it by the volume (area) of M.

Now, however, while the action is conformally invariant, the Laplace operator (1.1.103), (2.4.4)

$$\Delta = \frac{1}{\sqrt{g}} \frac{\partial^2}{\partial z \partial \bar{z}}$$

and therefore also its eigenvalues depend on the metric g and not only on its conformal class. When we consider a variation $g(x) \mapsto e^{\sigma(x)} g(x)$ of the metric, we can compute

$$\frac{\delta}{\delta \sigma(x)} \log \left(\frac{\det'(-\Delta)}{\mathrm{Vol}(M)} \right) \Big|_{\sigma=0} = -\frac{1}{12\pi} R(x), \tag{2.6.5}$$

where $R(x)$ is the scalar curvature of g, see e.g. [38], p. 145ff.

Denoting the partition function for the metric g by Z_g, we then have

$$\frac{\delta}{\delta \sigma(x)} Z_{e^{\sigma} g} |_{\sigma=0} = \frac{c}{24\pi} R(x) Z_g. \tag{2.6.6}$$

More generally, one defines the energy–momentum tensor as an operator by (2.5.71), that is,

$$\langle T_{\alpha_1 \beta_1}(z_1, \bar{z}_1) \cdots T_{\alpha_m \beta_m}(z_m, \bar{z}_m) \varphi(w_1, \bar{w}_1) \cdots \varphi(w_n, \bar{w}_n) \rangle_g$$

$$= \frac{1}{Z_g} \frac{(4\pi)^m \delta^m}{\delta g^{\alpha_1 \beta_1}(z_1, \bar{z}_1) \cdots \delta g^{\alpha_m \beta_m}(z_m, \bar{z}_m)} \left(Z_g \langle \varphi(w_1, \bar{w}_n) \cdots \varphi(w_n, \bar{w}_n) \rangle_g \right). \tag{2.6.7}$$

Thus, as an operator, the energy–momentum tensor takes into account the variation of the action S and of the integration measure $D\varphi$, as at the end of Sect. 2.5.3.

In particular

$$\langle T_{\alpha\beta}(z, \bar{z}) \rangle = \frac{4\pi}{Z_g} \frac{\delta}{\delta g^{\alpha\beta}(z, \bar{z})} Z_g. \tag{2.6.8}$$

At a conformal metric $g = \rho^2 |dz|^2$, that is, $g^{z\bar{z}} = 2\rho^{-2}$, $g^{zz} = 0 = g^{\bar{z}\bar{z}}$, we consider the above variation $g \mapsto e^{\sigma} g$ and obtain

$$\frac{4\pi}{Z_g} \frac{\delta}{\delta\sigma} Z_{e^{\sigma} g} |_{\sigma=0} = -g^{zz} \langle T_{zz} \rangle_g - 2g^{z\bar{z}} \langle T_{z\bar{z}} \rangle_g - g^{\bar{z}\bar{z}} \langle T_{\bar{z}\bar{z}} \rangle_g = -4\rho^{-2} \langle T_{z\bar{z}} \rangle_g. \tag{2.6.9}$$

From (2.6.8), (2.6.6), we then conclude

$$4\rho^{-2} \langle T_{z\bar{z}} \rangle_g = -\frac{c}{6} R. \tag{2.6.10}$$

Since the Euclidean metric ($g^{z\bar{z}} = 2$, $g^{zz} = 0 = g^{\bar{z}\bar{z}}$) has vanishing scalar curvature, we have there that

$$\langle T_{z\bar{z}} \rangle = 0, \tag{2.6.11}$$

that is, the energy–momentum tensor is traceless when the metric is Euclidean. For nonvanishing curvature R, however, T is no longer traceless. The trace given by (2.6.10) involves both the curvature and the central charge c.

From Axiom (ii), we obtain

$$\langle T_{zz}\rangle_{e^\sigma g} = \langle T_{zz}\rangle_g + \frac{c}{24}\frac{\delta}{\delta g^{zz}}\left(\|d\sigma\|^2_{L^2_g} + 4\int\sigma R\right)$$

$$= \langle T_{zz}\rangle_g - \frac{c}{12}\left(\partial_z^2\sigma - \frac{1}{2}(\partial_z\sigma)^2\right), \qquad (2.6.12)$$

using, for the last step, (2.4.6) and the formula

$$R = -\frac{1}{2}\left(\frac{\partial^2 g^{zz}}{\partial z^2} + \frac{\partial^2 g^{\bar z\bar z}}{\partial\bar z^2}\right) + \text{higher-order terms in } g^{zz}, g^{\bar z\bar z}, \qquad (2.6.13)$$

which is valid when we vary the Euclidean metric, that is, when we have $g^{z\bar z} = 2$ (see (1.1.148)). From (2.6.1) and (2.6.12), under a holomorphic transformation $z \mapsto w = f(z)$,

$$(f'(z))^2\langle T_{ww}\rangle_{dw\,d\bar w} = \langle T_{zz}\rangle_{|f'(z)|^2 dz\,d\bar z}$$

$$= \langle T_{zz}\rangle - \frac{c}{12}\left(\frac{\partial^2}{\partial z^2}\log f'(z) - \frac{1}{2}\left(\frac{\partial}{\partial z}\log f'(z)\right)^2\right)$$

$$= \langle T_{zz}\rangle - \frac{c}{12}\left(\frac{f'''(z)}{f'(z)} - \frac{3}{2}\left(\frac{f''(z)}{f'(z)}\right)^2\right)$$

$$= \langle T_{zz}\rangle - \frac{c}{12}\{f; z\}, \qquad (2.6.14)$$

where $\{f; z\}$ is the so-called Schwarzian derivative of f. So, we see here an important difference between the classical and the quantum energy–momentum tensor. While the latter is trace-free (2.6.11) for the Euclidean metric (but not in general) and holomorphic (2.6.17) (below) like the former, it no longer transforms as a quadratic differential, but instead picks up an additional term in its transformation rule (2.6.14). That term depends on the central charge c of the theory.

In order to take also variations w.r.t. g^{zz}, we now reconsider (2.6.9), (2.6.10) as

$$-g^{zz}\langle T_{zz}\rangle_g - 2g^{z\bar z}\langle T_{z\bar z}\rangle_g - g^{\bar z\bar z}\langle T_{\bar z\bar z}\rangle_g = \frac{c}{6}R. \qquad (2.6.15)$$

Next, applying $\frac{4\pi}{Z_g}\frac{\delta}{\delta g^{z_1z_1}}Z_g$ to (2.6.15) and recalling that the background metric is flat, that is, $g^{z\bar z} = 2$, $g^{zz} = 0 = g^{\bar z\bar z}$, as well as (2.6.13), and using (2.1.54), we obtain

$$4\pi\delta(z - z_1)\langle T_{zz}\rangle + 4\pi\langle T_{z_1z_1}T_{z\bar z}\rangle = \frac{\pi c}{3}\partial_z^2\delta(z - z_1). \qquad (2.6.16)$$

Next, diffeomorphism invariance implies that $\langle T_{zz} \rangle$ is holomorphic, as in the classical case,

$$\partial_{\bar{z}} \langle T_{zz} \rangle = 0 = \partial_z \langle T_{\bar{z}\bar{z}} \rangle. \tag{2.6.17}$$

Finally, one has the OPE

$$\langle T_{zz} T_{z_1 z_1} \rangle = \frac{c}{2(z-z_1)^4} + \frac{2}{(z-z_1)^2} \langle T_{z_1 z_1} \rangle$$

$$+ \frac{1}{z-z_1} \partial_{z_1} \langle T_{z_1 z_1} \rangle + \text{analytic terms in } z. \tag{2.6.18}$$

We shall now explain this in more detail.

2.6.2 Operator Product Expansions and the Virasoro Algebra

We take up the discussion of Sect. 2.5.3. As before in (2.5.58), we consider $z \mapsto z + \varepsilon f(z)$, f holomorphic.

We apply the general Ward identity (2.5.35) for $j = fT$, T being the energy–momentum tensor in CFT, writing $T(z)$ for T_{zz},

$$\frac{\partial}{\partial \bar{z}} \langle f T(z) \varphi(z_1) \cdots \varphi(z_n) \rangle = \sum_{k=1}^{n} \langle \varphi(z_1) \cdots \delta\varphi(z_k) \cdots \varphi(z_n) \rangle \delta(z - z_k) \tag{2.6.19}$$

to primary fields with variation (see (2.5.60))

$$\delta\varphi = h \partial_z f \varphi + f \partial_z \varphi. \tag{2.6.20}$$

We also write

$$\delta(z - z_k) = -\frac{1}{\pi} \partial_{\bar{z}} \left(\frac{1}{z - z_k} \right)$$

and integrate $\partial_z f \delta(z - z_k)$ by parts to obtain, using that (2.6.19) holds for all (holomorphic) f, and neglecting the factor π,

$$\frac{\partial}{\partial \bar{z}} \langle T(z) \varphi(z_1) \cdots \varphi(z_n) \rangle - \sum_{k=1}^{n} \left(\frac{1}{z - z_k} \partial_{z_k} + \frac{h}{(z - z_k)^2} \right) \langle \varphi(z_1) \cdots \varphi(z_n) \rangle = 0. \tag{2.6.21}$$

Under a holomorphic field f, T has to transform as

$$\delta_f T(z) = f(z) \partial_z T + 2(\partial_z f) T + \frac{c}{12} \partial_z^3 f, \tag{2.6.22}$$

because f transforms like $\frac{\partial}{\partial z}$, and T transforms like $(dz)^2$. As always, c is the central charge.

When we want to use $\varphi(z_1) = T(z_1)$ in the preceding, we therefore have to replace (2.6.20) by (2.6.22) and obtain (2.5.63), that is,

$$\langle T(z)T(z_1)\rangle = \frac{1}{z-z_1}\partial_{z_1}\langle T(z_1)\rangle + \frac{2}{(z-z_1)^2}\langle T(z_1)\rangle$$

$$+ \frac{c}{2}\frac{1}{(z-z_1)^4} + \text{analytic terms.} \qquad (2.6.23)$$

This is (2.6.18).

We also recall (2.5.57), saying that the transformation $z \mapsto z + f(z)$ is generated by

$$Q_f = \oint_C \frac{dz}{2\pi i} f(z)T(z).$$

Therefore, in particular,

$$[Q_f, T(w)] = f\partial_w T + 2(\partial_w f)T + \frac{c}{12}\partial_w^3 f.$$

As above, the commutator means that

$$\langle [Q_f, T(w)]\varphi(z_1)\cdots\varphi(z_n)\rangle$$

$$= \left(\oint_{C_1} \frac{dz}{2\pi i} - \oint_{C_2} \frac{dz}{2\pi i}\right) f(z)\langle T(z)T(w)\varphi(z_1)\cdots\varphi(z_n)\rangle,$$

where z lies inside C_1, but outside of C_2, while the z_k all lie inside C_2.

Integrating this with some function f_2 around the loop C_2 then leads to

$$[Q_{f_1}, Q_{f_2}] = Q_{[f_1, f_2]} + \frac{c}{24}\oint_C \frac{dz}{2\pi i}((\partial_z^3 f_1)f_2 - f_1\partial_z^3 f_2).$$

This then gives us the Virasoro algebra (2.5.66).

2.6.3 Superfields

We recall the basic transformation rules for a family of super Riemann surfaces from Sect. 1.5.3:

$$\tilde{z} = f(z) + \theta k(z),$$

$$\tilde{\vartheta} = g(z) + \theta h(z), \qquad (2.6.24)$$

$$f, g, k, h \text{ holomorphic}, \qquad \frac{\partial f}{\partial z} \neq 0.$$

We define

$$D_+ := \partial_\theta + \theta\partial_z, \qquad D_+^2 = \partial_z. \qquad (2.6.25)$$

(D_+ had been called τ in Sect. 1.5.3, and later on, we shall sometimes write θ_+ in place of θ, and θ_- in place of $\bar{\theta}$.)

The transformation law under holomorphic coordinate changes is

$$D_+ = (D_+\tilde{\theta})\widetilde{D_+} + (D_+\tilde{z} - \tilde{\theta}D_+\tilde{\theta})\widetilde{D_+}^2. \tag{2.6.26}$$

Superconformal means homogeneous transformation law, i.e.,

$$D_+\tilde{z} = \tilde{\theta}D_+\tilde{\theta}. \tag{2.6.27}$$

This is equivalent to

$$\begin{aligned}
\tilde{z} &= f(z) + \theta g(z)h(z), \\
\tilde{\theta} &= g(z) + \theta h(z)
\end{aligned} \tag{2.6.28}$$

with

$$h^2(z) = \frac{\partial f}{\partial z} + g(z)\frac{\partial g}{\partial z} \quad (g \text{ anticommuting}). \tag{2.6.29}$$

Since $D_+^2 = \partial_z$, (2.6.27) yields

$$\partial_z\tilde{z} + \tilde{\theta}\partial_z\tilde{\theta} = (D_+\tilde{\theta})^2 \tag{2.6.30}$$

as a compact version of the superconformal coordinate transformation rule.

In global terms, θ is a section of $K^{\frac{1}{2}}$, a square root of the canonical bundle of the underlying Riemann surface Σ. Such a square root of K corresponds to the choice of a spin structure on Σ. (To see this transformation behavior, put for example $g = 0$ in (2.6.28). Then from (2.6.29), $\tilde{\theta} = \sqrt{\frac{\partial f}{\partial z}}\theta$.)

We now look at the transformation behavior of conformal (primary) superfields

$$X(z, \theta) = \varphi(z) + \theta\psi(z)$$

of conformal weight h. Since θ as a section of $K^{\frac{1}{2}}$ has conformal weight $\frac{1}{2}$, this means that ψ has weight $h - \frac{1}{2}$, while φ has weight h. According to (2.6.30), we can also express the transformation law as

$$X(z, \theta) = X(\tilde{z}, \tilde{\theta})(D_+\tilde{\theta})^{2h} \tag{2.6.31}$$

(since $\tilde{\theta}$ has weight $\frac{1}{2}$ and D_+ has weight $-\frac{1}{2}$, $D_+\tilde{\theta}$ has weight 0, which it should, to make the transformation law consistent). Similarly, (2.5.60) becomes

$$\begin{aligned}
\delta_f\varphi(z) &= (h\partial_z f + f\partial_z)\varphi(z), \\
\delta_f\psi(z) &= \left(\left(h - \frac{1}{2}\right)\partial_z f + f\partial_z\right)\psi(z).
\end{aligned} \tag{2.6.32}$$

We also have the supersymmetry transformations for an anticommuting holomorphic g,

$$\delta_g \varphi(z) = \frac{1}{2} g \psi, \tag{2.6.33}$$

$$\delta_g \psi(z) = \frac{1}{2} g \partial_z \varphi + h \partial_z g \varphi, \tag{2.6.34}$$

that is,

$$\delta_g X(z, \theta) = \left(\frac{1}{2} g D_+ + h \partial_z g \right) X. \tag{2.6.35}$$

As before, we write this as a commutator with a charge

$$\delta_g X = -[Q_g, X] = \oint_c \frac{dz_1}{2\pi i} g(z_1) T(z_1) X(z) \tag{2.6.36}$$

for a small circle c about z.

Here, T is the (anticommuting) generator of the superconformal algebra. From this, we can draw the same consequences as above. We observe that for two supersymmetry transformations generated by g_1, g_2, if we put

$$f := \frac{1}{2} g_1 g_2, \tag{2.6.37}$$

we have

$$[\delta_{g_1}, \delta_{g_2}]_+ X(z, \theta) = \delta_f X(z, \theta), \tag{2.6.38}$$

where $+$ denotes the anticommutator. Thus, **a supersymmetry transformation is a square root of a conformal transformation**, as it should be according to (2.6.30).

As in Sect. 2.4.3, with $\psi_+ = \psi_1 - i\psi_2$, $\psi_- = \psi_1 + i\psi_2$ and $\theta_+ = \theta_1 + i\theta_2$, $\theta_- = \theta_1 - i\theta_2$ (alternatively, if we wished to conform to the notation in (2.6.24), we could write $\theta, \bar{\theta}$ in place of θ_+, θ_-), we use the operators $D_+ = \partial_{\theta_+} + \theta_+ \partial_z$, $D_- = \partial_{\theta_-} + \theta_- \partial_{\bar{z}}$ and consider a superfield

$$X = \phi + \frac{1}{2} (\psi_+ \theta_+ + \psi_- \theta_-) + \frac{i}{2} F \theta_+ \theta_- \tag{2.6.39}$$

and obtain the action

$$S = \int \frac{1}{2} D_- X D_+ X d^2 x d\theta_- d\theta_+$$

$$= \int \frac{1}{2} \left(4 \partial_z \phi \partial_{\bar{z}} \phi - \psi_+ \frac{\partial}{\partial \bar{z}} \psi_+ - \psi_- \frac{\partial}{\partial z} \psi_- + F^2 \right) d^2 z. \tag{2.6.40}$$

In Sect. 2.4.3, we derived the equations of motion (2.4.62),

$$D_- D_+ X = 0. \tag{2.6.41}$$

A solution can be decomposed as

$$X(z, \theta_+, \bar{z}, \theta_-) = X(z, \theta_+) + X(\bar{z}, \theta_-), \tag{2.6.42}$$

and we may write

$$X(z, \theta_+) = \varphi(z) + \theta_+ \psi_+(z). \tag{2.6.43}$$

The action is invariant under superconformal transformations and the corresponding energy–momentum tensor is

$$T = -\frac{1}{2} D_+ X \partial_z X = T_F + \theta_+ T_B, \tag{2.6.44}$$

with

$$T_F = -\frac{1}{2} \psi \partial_z \varphi, \tag{2.6.45}$$

$$T_B = -\frac{1}{2} (\partial_z \varphi)^2 - \frac{1}{2} \partial_z \psi \cdot \psi. \tag{2.6.46}$$

T_B is a section of K^2, T_F one of $K^{\frac{3}{2}}$.

We consider a complex Weyl spinor ψ_+ on a Riemann surface Σ, that is, a section of a spin bundle $K^{\frac{1}{2}}$, a square root of the canonical bundle K, given by a spin structure on Σ. We let ψ_- be the complex conjugate of ψ_+. Thus, ψ_- is a section of $\bar{K}^{\frac{1}{2}}$ (for the same spin structure).

We now consider the case where Σ is a cylinder, with coordinates $w = \tau + i\sigma$, identifying $\sigma + 2\pi$ with σ and with τ in some interval which is not further specified here. As there are two different spin structures on a cylinder, we have two choices for identifying ψ at $\sigma + 2\pi$ with ψ at σ:

$$\psi_\pm(\tau, \sigma + 2\pi) = \psi_\pm(\tau, \sigma), \quad \text{periodic (Ramond), or}$$
$$\psi_\pm(\tau, \sigma + 2\pi) = -\psi_\pm(\tau, \sigma), \quad \text{antiperiodic (Neveu–Schwarz).} \tag{2.6.47}$$

These boundary conditions also arise from the following consideration. We consider the half cylinder where σ runs from 0 to π, and we assume boundary relations between the holomorphic field ψ_+ and the antiholomorphic field ψ_-,

$$\psi_+(0, \tau) = \nu \psi_-(0, \tau) \quad \text{with } \nu = \pm 1,$$
$$\psi_+(\pi, \tau) = \psi_-(\pi, \tau) \tag{2.6.48}$$

where the factor $+1$ has been chosen w.l.o.g. in the second equation. We can then combine ψ_+ and psi_- into a single field, defined for $\sigma \in [0, 2\pi]$, by putting

$$\psi_+(\sigma, \tau) = \psi_-(2\pi - \sigma, \tau) \quad \text{for } \pi \le \sigma \le 2\pi. \tag{2.6.49}$$

ψ_+ then is holomorphic, because ψ_- was antiholomorphic. Also,

$$\psi_+(2\pi, \tau) = \psi_-(0, \tau) = \begin{cases} \psi_+(0, \tau) & \text{for } \nu = 1, \\ -\psi_+(0, \tau) & \text{for } \nu = -1. \end{cases} \tag{2.6.50}$$

Thus, ψ_+ is periodic (Ramond) in the first and antiperiodic (Neveu–Schwarz) in the second case.

We now map the cylinder to an annulus via

$$z = e^w.$$

Since ψ_+ transforms like $(dw)^{\frac{1}{2}}$, we have

$$\psi_+^{\text{annulus}}(z)(dz)^{\frac{1}{2}} = \psi_+^{\text{cylinder}}(w)(dw)^{\frac{1}{2}},$$

with

$$\left(\frac{dz}{dw}\right)^{\frac{1}{2}} = e^{\frac{w}{2}}.$$

When we now rotate the cylinder by 2π, the factor $e^{\frac{w}{2}}$ changes by a factor -1. Therefore, periodic and antiperiodic identifications are exchanged, and on the annulus, we have

$$\text{Ramond:} \quad \psi_\pm(e^{2\pi i} z) = -\psi_\pm(z) \quad \text{(antiperiodic)},$$
$$\text{Neveu–Schwarz:} \quad \psi_\pm(e^{2\pi i} z) = \psi_\pm(z) \quad \text{(periodic)}.$$

We shall now expand these expressions in terms of

$$z_{12} = z_1 - z_2 - \theta_1\theta_2,$$
$$\theta_{12} = \theta_1 - \theta_2.$$

We obtain

$$T(z_1, \theta_1)X(z_2, \theta_2) = h\frac{\theta_{12}}{z_{12}^2}X(z_2, \theta_2) + \frac{1}{2z_{12}}D_{+,2}X(z_2, \theta_2)$$

$$+ \frac{\theta_{12}}{z_{12}}\partial_{z_2}X(z_2, \theta_2) + \text{regular terms},$$

$$T(z_1, \theta_1)T(z_2, \theta_2) = \frac{c}{6}\frac{1}{z_{12}^3} + \frac{3}{2}\frac{\theta_{12}}{z_{12}^2}T(z_2, \theta_2) + \frac{1}{2z_{12}}D_{+,2}T(z_2, \theta_2)$$

$$+ \frac{\theta_{12}}{z_{12}}\partial_{z_2}T(z_2, \theta_2) + \text{regular terms}.$$

In components:

$$T_B(z_1)T_B(z_2) = \frac{c}{6}\frac{1}{(z_1-z_2)^4} + \frac{2}{(z_1-z_2)^2}T_B(z_2) + \frac{1}{z_1-z_2}\partial_{z_2}T_B(z_2) + \cdots,$$

$$T_B(z_1)T_F(z_2) = \frac{3}{2}\frac{1}{(z_1-z_2)^2}T_F(z_2) + \frac{1}{z_1-z_2}\partial_{z_2}T_F(z_2) + \cdots,$$

$$T_F(z_1)T_F(z_2) = \frac{c}{6}\frac{1}{(z_1-z_2)^3} + \frac{1}{2}\frac{1}{z_1-z_2}T_B(z_2) + \cdots.$$

We expand T_B as before and T_F as

$$T_F(z) = \frac{1}{2}\sum_{k\in\mathbb{Z}+a} z^{-k-1-a}G_k \quad \left(G_k = 2\oint_c \frac{dz_i}{2\pi i}T_F(z)z^{k+a}\right),$$

with $a = 0$ corresponding to the Ramond sector and $a = \frac{1}{2}$ corresponding to the Neveu–Schwarz sector.

With $\hat{c} = \frac{2}{3}c$, we obtain the super Virasoro algebra

$$[L_m, L_n]_- = (m-n)L_{m+n} + \frac{\hat{c}}{8}(m^3 - m)\delta_{m+n},$$

$$[L_m, G_k]_- = \left(\frac{1}{2}m - k\right)G_{m+k},$$

$$[G_k, G_l]_+ = 2L_{k+l} + \frac{\hat{c}}{2}\left(k^2 - \frac{1}{4}\right)\delta_{k+l}.$$

2.7 String Theory

In conformal field theory, Sect. 2.6, we have kept the Riemann surface Σ fixed and varied the metric on Σ only via diffeomorphisms—which left the partition and correlation functions invariant—and by conformal changes—which, in contrast to the classical case, had a nontrivial effect, the so-called conformal anomaly. In string theory, one also varies the Riemann surface Σ itself. Equivalently, as explained in 7 in Sect. 1.4.2, we permit any variation of the metric γ, including those that change the underlying conformal structure. Here, we can only give some glimpses of the theory. Fuller treatments are given in [50, 77, 87, 88] and, closest to the presentation here, in [62].

In bosonic string theory, one starts with the linear sigma model (Polyakov action) (2.4.7)

$$S(\varphi, \gamma) \tag{2.7.1}$$

and considers the functional integral

$$Z = \sum_{\text{topological types}} \int e^{-S(\varphi,\gamma)}d\varphi d\gamma. \tag{2.7.2}$$

This means that one wishes to average over all fields ϕ and all compact[20] surfaces, described by their topological type (their genus) and their metric, with exponential weight coming from the Polyakov action. Since, as discussed, that action $S(\varphi, \gamma)$ is invariant under diffeomorphisms and conformal changes, that is, possesses an infinite-dimensional invariance group, this functional integral, as it stands, can only be infinite itself. Therefore, one divides out these invariances before performing the functional integral. As described in Sect. 1.4.2, the remaining degrees of freedom are the ones coming from the moduli of the underlying surface, and we are left with an integral over the Riemann moduli space for surfaces of given genus and a sum over all genera. The essential mathematical content of string theory is then to define that integral in precise mathematical terms and try to evaluate it. The sum needs some regularization, that is, one should put in some factor κ_p depending on the genus p that goes to 0 in some appropriate manner as the genus increases. Alternatively, one should construct a common moduli space that simultaneously includes surfaces of all genera. Since lower-genus surfaces occur in the compactification of the moduli spaces of higher-genus ones, this seems reasonable. As discussed above in Sect. 1.4.2, however, the Mumford–Deligne compactification is not directly appropriate for this, as there the lower-genus surfaces that occur in the boundary of the moduli space carry marked points in addition. With each reduction of the genus, the number of those marked points increases by two. When we then consider surfaces of some fixed genus p_0 in a boundary stratum of the moduli space of surfaces of genus p, we have $2(p - p_0)$ marked points, and this number then tends to ∞ for $p \to \infty$. Therefore, we need to resort to the Satake–Baily compactification described in Sect. 1.4.2 which does not need marked points, but is highly singular. We also recall from there that this compactification can be mapped into the Satake compactification of the moduli space of principally polarized Abelian varieties. Again, the compactification of that moduli space for principally polarized Abelian varieties of dimension p contains in its boundary the moduli spaces for the Abelian varieties of smaller dimension. Letting $p \to \infty$ then gives some kind of universal moduli space for principally polarized Abelian varieties of finite dimension, and this space is then stratified according to dimension. Similarly, the analogous universal moduli space for compact Riemann surfaces would then be stratified according to genus. (To the author's knowledge, however, this construction has never been carried through in detail.)

In any case, even the integral over the moduli space for a fixed genus leads to some subtleties. The reason is that while the Polyakov action $S(\varphi, \gamma)$ itself is conformally invariant, the measure $e^{-S(\varphi, \gamma)} d\phi d\gamma$ in (2.7.2) is not. We have seen the reason above from a somewhat different perspective in our discussion of quantization of the sigma model, where we encountered additional terms in the operator expansions. These then led to the nontrivial central charge c of the Virasoro algebra. It then turns out that there are two different sources of this conformal anomaly, one coming from the fields ϕ and the other from the metric γ. The fields are mappings

---

[20]Since the partition function represents the amplitude of vacuum $\to$ vacuum transitions, only closed surfaces are taken into account.

into some euclidean space $\mathbb{R}^d$, and we get a contribution to the conformal anomaly for each dimension, that is, an overall contribution proportional to d. The conformal anomaly coming from γ is independent of the target dimension d. It then turns out that these two conformal anomalies cancel precisely in dimension $d = 26$. Mathematically, this can be explained in terms of the geometry of the Riemann moduli space, utilizing earlier work of Mumford [84], or with the help of the semi-infinite cohomology of the Virasoro algebra. In conclusion, bosonic string theory lives in a 26-dimensional space.

The same scheme applies in superstring theory. Here, the action is given by (2.4.149),

$$S(\phi, \psi, \gamma, \chi) = \frac{1}{2} \int_\Sigma (\gamma^{\alpha\beta} \partial_\alpha \phi^a \partial_\beta \phi_a + \bar{\psi}^a \gamma^\alpha \partial_\alpha \psi_a + 2\bar{\chi}_\alpha \gamma^\beta \gamma^\alpha \psi^a \partial_\beta \phi_a$$
$$+ \frac{1}{2} \bar{\psi}_a \psi^a \bar{\chi}_\alpha \gamma^\beta \gamma^\alpha \chi_\beta) \sqrt{\det\gamma} \, dz^1 dz^2, \tag{2.7.3}$$

including also the fermionic field ψ and the gravitino χ. The same quantization principle is applied, and the resulting dimension needed to cancel the conformal anomalies turns out to be $d = 10$.

In order to include gravitational fields, one has to consider more general targets than euclidean space. The appropriate target spaces are Kähler manifolds with vanishing Ricci curvature. The real dimension still has to be 10. In order to make contact with dimension 4 of ordinary space–time, one writes such a target as a product

$$\mathbb{R}^4 \times M \tag{2.7.4}$$

where M now is assumed to be compact (and of such a small scale that it is not directly observable at the macroscopic level). (This vindicates the old idea of Kaluza described in Sect. 1.2.4 above.) The process of making some of the dimensions compact is called compactification in the physics literature. M then has to be a compact Kähler manifold with vanishing Ricci curvature, in order to obtain supersymmetry, of complex dimension 3, a Calabi–Yau space. In fact, by Yau's theorem [109], every compact Kähler manifold with vanishing first Chern class $c_1(M)$ carries such a Ricci flat metric, and this makes the methods of algebraic geometry available for the investigation and classification of such spaces.

In order to describe the physical content of string theory, the basic object is the string, an open or closed curve. As it moves in space–time, it sweeps out a Riemann surface. In contrast to the mathematical framework just described, this Riemann surface will have boundaries, even in the case of a closed string when we follow it between two different times t_1 and t_2. The boundaries will then correspond to the initial position at time t_1 and the final position at time t_2, except when the string only comes into existence after time t_1 and ceases to exist at time t_2. See [62] for the systematic treatment of such boundaries in string theory. For an open string, that is, for a curve with two endpoints moving in space–time, we obtain further boundaries corresponding to the trajectories of these endpoints. More generally, the movement of these endpoints may be confined to lower-dimensional objects in space–time that carry charges and that can then become objects in their own right, the D-branes[21] first introduced by Polchinski [86]. Symmetries between branes then led to a new relation between string theory and gauge theory, culminating in a conjecture of Maldacena [78].

In any case, when a string moves in space–time, it sweeps out a surface, and the basic Nambu–Goto action of string theory was the area of that surface. Since the area functional is invariant under any reparametrization, it cannot be readily quantized, and therefore, the symmetry was reduced by considering the map that embeds the surface representing the moving string into space–time and the underlying metric of that surface as independent variables of the theory. That led to the Polyakov action (2.7.1), that is, the Dirichlet integral or sigma model action (2.4.7).

According to string theory, all elementary particles are given by vibrations of strings. Gauge fields arise from vibrations of open strings. Their endpoints represent charged particles. For instance, when one is an electron and the other an oppositely charged particle, a positron, the massless vibration of the string connecting them represents a photon that carries the electrical force between them. Collisions between such particles then naturally lead to closed strings. Gravitons, that is, particles responsible for the effects of gravity, arise from vibrations of closed strings. In superstring theory, both bosons and fermions are oscillations of strings. There are only two fundamental constants in string theory, in contrast to the proliferation of such constants in the standard model. These are the string tension, that is, the energy per unit-length of a string, the latter given in terms of the Planck length, and the string coupling constant, the probability for a string to break up into two pieces.

However, superstring theory is far from being unique, and it cannot determine the geometry of the background space–time purely on the basis of physical principles. Thus, there is room for further work in superstring theory, as well as for research on competing theories like loop quantum gravity (that started with Ashtekar's reformulation of Einstein's theory of general relativity [5]) and the development of new ones.

---

[21] The "D" here stands for Dirichlet, because such types of boundary conditions are called Dirichlet boundary conditions in the mathematical literature. We also recall that the basic action functional (2.4.7), (2.7.1) is called the Dirichlet integral in the mathematical literature. This terminology was in fact introduced by Riemann when he systematically used variational principles in his theory of Riemann surfaces, see [91]. Harmonic functions are minimizers of the Dirichlet integral, and in this sense, string theory is a quantization of the profound ideas of Riemann.

Bibliography

1. C. Albertsson, U. Lindström, and M. Zabzine. $N = 1$ supersymmetric sigma model with boundaries, I. arxiv:hep-th/0111161v4, 2003.
2. S. Albeverio and R. Høegh-Krohn. *Mathematical theory of Feynman path integrals*. Lecture notes in mathematics, volume 523. Springer, Berlin, 1976.
3. L. Alvarez-Gaumé. Supersymmetry and the Atiyah-Singer index theorem. *Commun. Math. Phys.*, 90:161–173, 1983.
4. S. Arakelov. Intersection theory of divisors on an arithmetic surface. *Izv. Akad. Nauk*, 38:1179–1192, 1974.
5. A. Ashtekar. New variables for classical and quantum gravity. *Phys. Rev. Lett.*, 57:2244–2247, 1986.
6. M. Atiyah, R. Bott, and A. Shapiro. Clifford modules. *Topology*, 3(Suppl. I):3–38, 1964.
7. M. Atiyah and I. Singer. The index of elliptic operators, III. *Ann. Math.*, 87:546–604, 1968.
8. W. Baily. On the moduli of Jacobian varieties. *Ann. Math.*, 71:303–314, 1960.
9. A.A. Belavin, A.M. Polyakov, and A.B. Zamolodchikov. Infinite conformal symmetry in two-dimensional quantum field theory. *Nucl. Phys. B*, 241:333–380, 1984.
10. N. Berline, E. Getzler, and M. Vergne. *Heat kernels and Dirac operators*. Springer, Berlin, 1992.
11. J. Bernstein, D. Leites, and V. Shander. *Supersymmetries: algebra and calculus*. 2006.
12. L. Bers. Spaces of degenerating surfaces. In *Discontinuous groups and Riemann surfaces*. Annals of mathematics studies, volume 79. Princeton University Press, Princeton, 1974.
13. F. Bethuel, H. Brezis, and F. Hélein. *Ginzburg-Landau vortices*. Birkhäuser, Basel, 1994.
14. G. Buss. Higher Bers maps. arXiv:0812.0314.
15. K.C. Chang. *Infinite dimensional Morse theory and multiple solution problems*. Birkhäuser, Basel, 1993.
16. Q. Chen, J. Jost, J.Y. Li, and G.F. Wang. Dirac-harmonic maps. *Math. Z.*, 254:409–432, 2006.
17. Q. Chen, J. Jost, and G.F. Wang. The supersymmetric non-linear σ-model. 2008.
18. T. Chinburg. An introduction to Arakelov intersection theory. In G. Cornell, J. Silverman, editors, *Arithmetic geometry*, pages 289–307. Springer, Berlin, 1986.
19. B. Clarke. The completion of the manifold of Riemannian metrics. arXiv:0904.0177.
20. B. Clarke. The metric geometry of the manifold of Riemannian metrics over a closed manifold. arXiv:0904.0174.
21. L. Crane and J. Rabin. Super Riemann surfaces: uniformization and Teichmüller theory. *Commun. Math. Phys.*, 4:601–623, 1988.
22. P. Deligne. Notes on spinors. In P. Deligne, et al., editors, *Quantum fields and strings: a course for mathematicians*, volume I, pages 99–135. Am. Math. Soc. and Inst. Adv. Study, Princeton, 1999.
23. P. Deligne and D. Freed. Supersolutions. In P. Deligne, et al., editors, *Quantum fields and strings: a course for mathematicians*, volume I, pages 227–355. Am. Math. Soc. and Inst. Adv. Study, Princeton, 1999.
24. P. Deligne and J. Morgan. Notes on supersymmetry (following Joseph Bernstein). In P. Deligne, et al., editors, *Quantum fields and strings: a course for mathematicians*, volume I, pages 41–97. Am. Math. Soc. and Inst. Adv. Study, Princeton, 1999.
25. P. Deligne and D. Mumford. The irreducibility of the space of curves of given genus. *Publ. Math. IHES*, 36:75–110, 1969.
26. E. D'Hoker. String theory. In P. Deligne, et al., editor, *Quantum fields and strings: a course for mathematicians*, volume 2, pages 807–1012. AMS, Providence, 1999.
27. E. D'Hoker and D.H. Phong. The geometry of string perturbation theory. *Rev. Mod. Phys.*, 60:917–1065, 1988.
28. W.Y. Ding, J. Jost, J.Y. Li, X.W. Peng, and G.F. Wang. Self duality equations for Ginzburg-Landau and Seiberg-Witten type functionals with 6th order potentials. *Commun. Math. Phys.*, 217(2):383–407, 2001.

29. W.Y. Ding, J. Jost, J.Y. Li, and G.F. Wang. An analysis of the two-vortex case in the Chern-Simons-Higgs model. *Calc. Var.*, 7:87–97, 1998.

30. W.Y. Ding, J. Jost, J.Y. Li, and G.F. Wang. Multiplicity results for the two vortex Chern-Simons-Higgs model on the two-sphere. *Comment. Math. Helv.*, 74:118–142, 1999.

31. S. Donaldson and P. Kronheimer. *The geometry of four-manifolds*. Oxford University Press, Oxford, 1990.

32. D. Leites, et al. *Superconformal surfaces and related structures*, 2006.

33. P. Deligne, et al.*Quantum fields and strings: a course for mathematicians*, volume I. Am. Math. Soc. and Inst. Adv. Study, Princeton, 1999.

34. P. Deligne, et al.*Quantum fields and strings: a course for mathematicians*, volume II. Am. Math. Soc. and Inst. Adv. Study, Princeton, 1999.

35. G. Faltings. Endlichkeitssätze für abelsche Varietäten über Zahlkörpern. *Inv. Math.*, 73:349–366, 1983.

36. G. Faltings. Calculus on arithmetic surfaces. *Ann. Math.*, 119:387–424, 1984.

37. A. Floer. Witten's complex and infinite dimensional Morse theory. *J. Differ. Geom.*, 30:207–221, 1989.

38. P. Di Francesco, P. Mathieu, and D. Sénéchal. *Conformal field theory*. Springer, Berlin, 1997.

39. D. Freed. *Five lectures on supersymmetry*. Am. Math. Soc., Providence, 1999.

40. D. Freed. Solution to problem FP2. In P. Deligne, et al., editor, *Quantum fields and strings: a course for mathematicians*, volume I, pages 649–656. Am. Math. Soc. and Inst. Adv. Study, Princeton, 1999.

41. D. Friedan and P. Windey. Supersymmetric derivation of the Atiyah-Singer index and the chiral anomaly. *Nucl. Phys. B*, 235:395–416, 1984.

42. D. Friedan and P. Windey. Supersymmetry and index theorems. *Physica D*, 15:71–74, 1985.

43. J. Fröhlich, O. Grandjean, and A. Recknagel. Supersymmetric quantum theory and differential geometry. *Commun. Math. Phys.*, 193:527–594, 1998.

44. J. Fröhlich and M. Struwe. Variational problems on vector bundles. *Commun. Math. Phys.*, 131:431–464, 1990.

45. W. Fulton and J. Harris. *Representation theory*. Springer, Berlin, 1991.

46. K. Gawędzki. Lectures on conformal field theory. In P. Deligne, et al., editors, *Quantum fields and strings: a course for mathematicians*, volume II, pages 727–805. Am. Math. Soc. and Inst. Adv. Study, Princeton, 1999.

47. E. Getzler. Pseudodifferential operators on manifolds and the index theorem. *Commun. Math. Phys.*, 92:163–178, 1983.

48. E. Getzler. A short proof of the Atiyah-Singer index theorem. *Topology*, 25:111–117, 1986.

49. J. Glimm and A. Jaffe. *Quantum Physics. A functional integral point of view*. Springer, Berlin, 1981.

50. M. Green, J. Schwarz, and E. Witten. *Superstring theory, I, II*. Cambridge University Press, Cambridge, 1995.

51. P. Griffiths and J. Harris. *Principles of algebraic geometry*. Wiley-Interscience, New York, 1978.

52. D. Gross. Renormalization groups. In P. Deligne, et al., editor, *Quantum fields and strings: a course for mathematicians*, volume I, pages 551–593. Am. Math. Soc. and Inst. Adv. Study, Princeton, 1999.

53. S. Gustafson and I. Sigal. *Mathematical concepts of quantum mechanics*. Springer, Berlin, 2003.

54. L. Habermann and J. Jost. Riemannian metrics on Teichmüller space. *Man. Math.*, 89:281–306, 1996.

55. L. Habermann and J. Jost. Metrics on Riemann surfaces and the geometry of moduli spaces. In J.P. Bourguignon, P. di Bartolomeis, and M. Giaquinta, editors, *Geometric theory of singular phenomena in partial differential equations*, Cortona 1995, pages 53–70. Cambridge University Press, Cambridge, 1998.

56. S. Hawking and Ellis. *The large scale structure of space–time*. Cambridge University Press, Cambridge, 1973.

57. B. Helffer and J. Sjostrand. Puits multiples en mécanique semi classique, IV: etude du complexe de Witten. *Commun. Partial Differ. Equ.*, 10:245–340, 1985.
58. Z. Huang. Calculus of variations and the L2-Bergman metric on Teichmüller space. arXiv:math.DG/0506569, 2006.
59. J. Jorgenson and J. Kramer. Non-completeness of the Arakelov-induced metric on moduli space of curves. *Man. Math.*, 119:453–463, 2006.
60. J. Jost. Orientable and nonorientable minimal surfaces. *Proc. First World Congress of Nonlinear Analysts, Tampa, Florida*, 1992:819–826, 1996.
61. J. Jost. Minimal surfaces and Teichmüller theory. In S.T. Yau, editor, *Tsing Hua lectures on geometry and analysis*, pages 149–211. International Press, Boston, 1997.
62. J. Jost. *Bosonic strings*. International Press, Boston, 2001.
63. J. Jost. *Partial differential equations*, 2nd edn. Springer, Berlin, 2007.
64. J. Jost. *Compact Riemann surfaces*, 3rd edn. Springer, Berlin, 2006.
65. J. Jost. *Riemannian geometry and geometric analysis*, 5th edn. Springer, Berlin, 2008.
66. J. Jost and X. Li-Jost. *Calculus of variations*. Cambridge University Press, Cambridge, 1998.
67. J. Jost and X.W. Peng. Group actions, gauge transformations and the calculus of variations. *Math. Ann.*, 293:595–621, 1992.
68. J. Jost and M. Struwe. Morse-Conley theory for minimal surfaces of varying topological type. *Inv. Math.*, 102:465–499, 1990.
69. J. Jost and S.T. Yau. Harmonic maps and algebraic varieties over function fields. *Am. J. Math.*, 115:1197–1227, 1993.
70. J. Jost and K. Zuo. Harmonic maps of infinite energy and rigidity results for Archimedean and non-Archimedean representations of fundamental groups of quasiprojective varieties. *J. Differ. Geom.*, 47:469–503, 1997.
71. J. Jost and K. Zuo. Harmonic maps into Bruhat-Tits buildings and factorizations of p-adically unbounded representations of π_1 of algebraic varieties, I. *J. Algebraic Geom.*, 9:1–42, 2000.
72. J. Jost and K. Zuo. Representations of fundamental groups of algebraic manifolds and their restrictions to fibers of a fibration. *Math. Res. Lett.*, 8:569–575, 2001.
73. S.V. Ketov. *Quantum non-linear sigma-models*. Springer, Berlin, 2000.
74. C. Kiefer. *Quantum gravity*. Oxford University Press, Oxford, 2007.
75. A. Knapp. *Representation theory of semisimple groups*, 2nd edn. Princeton University Press, Princeton, 2001.
76. D. Leites. Introduction to the theory of supermanifolds. *Usp. Mat. Nauk*, 35:3–57, 1980. Tanslated in: *Russ. Math. Surv.*, 35:1–64, 1980.
77. D. Lüst and S. Theisen. *Lectures on string theory*. Lecture notes in physics, volume 346. Springer, Berlin, 1989.
78. J. Maldacena. The large N limit of superconformal field theories. *Adv. Theor. Math. Phys.*, 2:231–252, 1998.
79. Y. Manin. *Gauge field theory and complex geometry*, 2nd edn. Springer, Berlin, 1997.
80. H. Masur. The extension of the Weil-Petersson metric to the boundary of Teichmüller space. *Duke Math. J.*, 43:623–635, 1977.
81. C. Misner, K. Thorne, and J. Wheeler. *Gravitation*. Freeman, New York, 1973.
82. L. Modica. The gradient theory of phase transitions and the minimal interface criterion. *Arch. Ration. Mech. Anal.*, 98:123–142, 1987.
83. D. Mumford. *Geometric invariant theory*. Springer, Berlin, 1965.
84. D. Mumford. Stability of projective varieties. *Enseign. Math.*, 23:39–110, 1977.
85. R. Penrose. *The road to reality*. Jonathan Cape, 2004.
86. J. Polchinski. Dirichlet branes and Ramond-Ramond charges. *Phys. Rev. Lett.*, 75:4724–4727, 1995.
87. J. Polchinski. *String theory*, volume I. Cambridge University Press, Cambridge, 1998.
88. J. Polchinski. *String theory*, volume II. Cambridge University Press, Cambridge, 1998.
89. J. Rabin. Super Riemann surfaces. In S.T. Yau, editor, *Mathematical aspects of string theory*, pages 368–385. World Scientific, Singapore, 1987.

90. J. Rabin. Introduction to quantum field theory for mathematicians. In D. Freed, K. Uhlenbeck, editors, *Geometry and quantum field theory*, pages 183–269. Am. Math. Soc., Providence, 1995.

91. B. Riemann. In R. Narasimhan, editor, *Gesammelte mathematische Werke, wissenschaftlicher Nachlass und Nachträge. Collected papers.* Springer, Berlin, 1990.

92. C. Rovelli. *Quantum gravity.* Cambridge University Press, Cambridge, 2004.

93. C. Sachse. *Global analytic approach to super Teichmüller spaces.* University of Leipzig, 2007. Also in: arXiv:0902.3389v1.

94. C. Sachse. A categorical formulation of superalgebra and supergeometry. arXiv:0802.4067, 2008.

95. M. Schlicht. Another proof of Bianchi's identity in arbitrary bundles. *Ann. Glob. Anal. Geom.*, 13:19–22, 1995.

96. M. Schwarz. *Morse homology.* Birkhäuser, Basel, 1993.

97. T. Shiota. Characterization of Jacobian varieties in terms of soliton equations. *Inv. Math.*, 83:333–386, 1986.

98. S. Sternberg. *Group theory and physics.* Cambridge University Press, Cambridge, 1994.

99. C. Taubes. Arbitrary n-vortex solutions to the first order Ginzburg-Landau equations. *Commun. Math. Phys.*, 72:227–292, 1980.

100. H. Triebel. *Höhere Analysis.* Harry Deutsch, Thun, 1980.

101. A. Tromba. *Teichmüller theory in Riemannian geometry.* Birkhäuser, Basel, 1992.

102. V.S. Varadarajan. *Supersymmetry for mathematicians: an introduction.* Courant Institute for Mathematical Sciences and American Mathematical Society, 2004.

103. St. Weinberg. *The quantum theory of fields,* volume I. Cambridge University Press, Cambridge, 1995.

104. E. Witten. Constraints on supersymmetry breaking. *Nucl. Phys. B*, 202:253–316, 1982.

105. E. Witten. Supersymmetry and Morse theory. *J. Differ. Geom*, 17:661–692, 1982.

106. E. Witten. Perturbative quantum field theory. In P. Deligne, et al., editor, *Quantum fields and strings: a course for mathematicians,* volume I, pages 419–473. Am. Math. Soc. and Inst. Adv. Study, Princeton, 1999.

107. S. Wolpert. Noncompleteness of the Weil-Petersson metric for Teichmüller space. *Pac. J. Math.*, 61:513–576, 1975.

108. S. Wolpert. Geometry of the Weil-Petersson completion of Teichmüller space. arXiv: math/0502528, 2005.

109. S.T. Yau. On the Ricci curvature of a compact Kähler manifold and the complex Monge-Ampère equation, I. *Commun. Pure Appl. Math.*, 31:339–411, 1978.

110. K. Yosida. *Functional analysis,* 5th edn. Springer, Berlin, 1978.

111. E. Zeidler. *Quantum field theory (6 vols.).* Springer, Berlin, 2006.

112. J. Zinn-Justin. *Path integrals in quantum mechanics.* Oxford University Press, Oxford, 2005.

113. J. Zinn-Justin. *Phase transitions and renormalization group.* Oxford University Press, Oxford, 2007.

114. J. Zinn-Justin. *Quantum field theory and critical phenomena,* 4th edn. Oxford University Press, Oxford, 2002.

Index

Abelian variety, 70, 77, 82
action, 21, 122
action principle, 97
adjoint of exterior derivative, 20
algebraic curve, 67
algebraic variety, 69
analytic continuation, 113
annihilation operator, 171
antiholomorphic, 14
antiself-dual, 6, 46
Arakelov metric, 75, 76
area form, 15
automorphism bundle, 42
autoparallel, 12, 13

Baily–Satake compactification, 82
baryon, 134
Berezin integral, 86, 159
Bergmann metric, 75
Bianchi identity, 27, 28, 41
boson, 133, 139
bosonic multiplet, 143
bosonic string theory, 204
brane, 163
broken symmetry group, 137

Calabi–Yau space, 206
canonical bundle, 68, 71
central charge, 193–195, 197, 198
charged particle, 123
Chern classes, 45
Chern form, 77
chiral representation, 62
chirality operator, 55
Christoffel symbols, 10, 13
Clifford algebra, 50, 123, 158, 170
closed form, 20
closing of supersymmetry algebra, 144, 161
coclosed form, 21
cohomology, 169
cohomology group, 20
color, 134
commutative super algebra, 83
commutator algebra of charges, 191
compactification, 206
compactification of moduli space, 78, 80–82, 205
complex Clifford algebra, 51
complex conjugation, 84

complex manifold, 16, 49
complex space, 13
complex super vector space, 84
complex tangent space, 16
complex wave function, 107
complexification, 14
conformal anomaly, 204, 205
conformal covariance, 195
conformal field theory, 195, 204
conformal invariance, 153, 157, 158, 179, 184, 190
conformal map, 157
conformal spin, 50
conformal structure, 71
conformal superfield, 200
conformal transformation, 192, 201
conformal weight, 50, 192
connection, 9, 37
conserved charge, 150, 192
conserved current, 155, 188, 191
contravariant, 5, 49
coordinate change, 2, 3
coordinate representation, 1
correlation function, 184, 195
correspondence between classical and quantum mechanics, 110
cosmological constant, 32
cotangent bundle, 49
cotangent space, 4, 49
cotangent vector, 4
coupling constant, 128, 129
covariant, 5, 49
covariant derivative, 9, 37
covariantly constant, 11
covector, 4
creation operator, 171
curvature form, 77
curvature of connection, 40
curvature tensor, 12, 26, 29

D-brane, 163
de Rham cohomology, 20, 171
degeneration of Riemann surface, 79–82
degree of Clifford algebra element, 51
degree of divisor, 68
degree of line bundle, 68
determinant, 184
diffeomorphism covariance, 195
diffeomorphism invariance, 155, 157, 198
differentiable manifold, 2